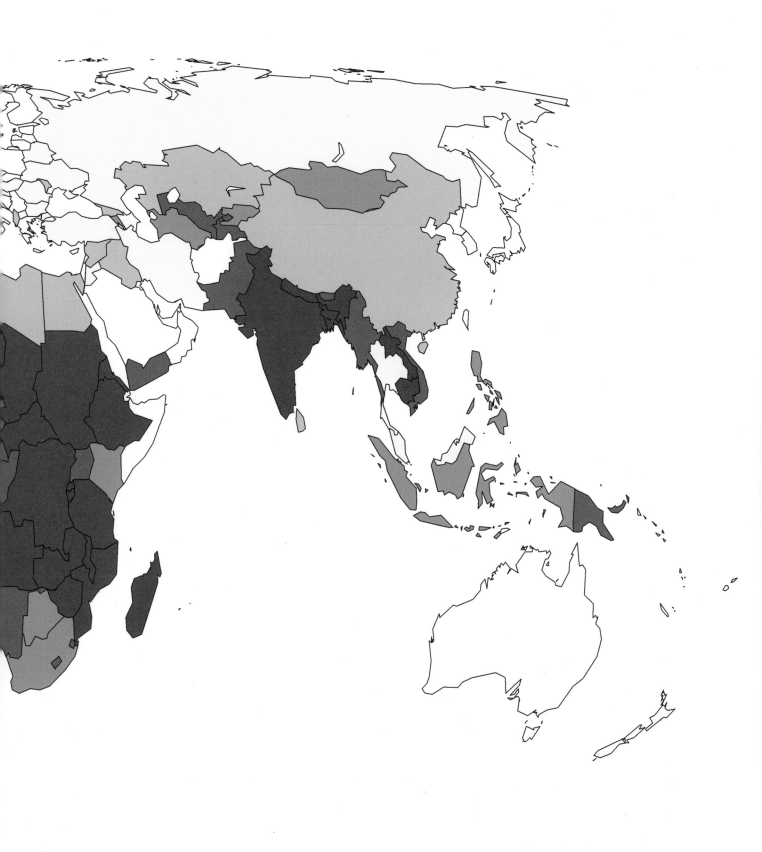

# The Atlas of World Hunger

# The Atlas of World Hunger

**Thomas J. Bassett and Alex Winter-Nelson**

*The University of Chicago Press*   Chicago and London

**Thomas J. Bassett** is professor of geography at the University of Illinois, Urbana-Champaign.
**Alex Winter-Nelson** is professor of agricultural and consumer economics at the University of Illinois, Urbana-Champaign.

The University of Chicago Press, Chicago 60637
The University of Chicago Press, Ltd., London
© 2010 by Thomas J. Bassett and Alex Winter-Nelson
All rights reserved. Published 2010
Printed in China

18 17 16 15 14 13 12 11 10     1 2 3 4 5

ISBN-13: 978-0-226-03907-7 (cloth)
ISBN-10: 0-226-03907-2 (cloth)

Frontispiece: Elementary school students in the Korhogo region of northern Côte d'Ivoire participating in a school-lunch program.

Library of Congress Cataloging-in-Publication Data

Bassett, Thomas J.
   The atlas of world hunger / Thomas J. Bassett and
Alex Winter-Nelson.
      p. cm.
   Includes bibliographical references and index.
   ISBN-13: 978-0-226-03907-7 (cloth : alk. paper)
   ISBN-10: 0-226-03907-2 (cloth : alk. paper)
   1. Food supply—Maps.  2. Atlases.  I. Winter-Nelson,
Alex E.  II. Title.
   G1046.E59B38 2010
   363.8022′3—dc22                    2009022152

For more information on world hunger, including resources for teachers, visit worldhungeratlas.org.

# Contents

List of Maps                                                        vi
List of Figures                                                     viii
List of Tables                                                       x
List of Boxes                                                       xi
Acknowledgments                                                    xiii

1.  Introduction                                                     1

**Part I. Locating Hunger**                                         11

Indicators of Malnutrition                                          13
2.  Food Availability                                               15
3.  Prevalence of Undernourishment (POU)                           19
4.  Micronutrient Malnutrition                                      26
5.  Malnutrition and Obesity                                        32
6.  Growth Failure                                                  37
7.  Household Poverty                                               43
8.  The Hunger Vulnerability Index                                  51

Patterns of Hunger within Countries                                57
9.  Child Growth Failure in Uruguay                                59
10. Food Insecurity in the United States                           63
11. Malnutrition in India and Mexico                               67

**Part II. The Sources of Hunger**                                 71

National Resources                                                 73
12. Population Growth                                               75
13. Arable Land per Capita                                          79
14. Environmental Systems Health                                    84
15. Human Resources: Literacy                                       86
16. Built Resources: Roads                                          88
17. Change in Resource Base                                         91
18. Climate Change                                                  94

Technology                                                         97
19. College and University Enrollment and
    Research and Development                                        99
20. Agricultural Technology: Fertilizer                            103

Institutions and Power Relations                                  107
21. Colonialism and Neocolonialism                                109
22. Debt and International Power Relations                         112
23. Political Freedoms                                             115
24. Income Inequality                                              118
25. Gender Inequality                                              122

Poverty                                                            125
26. National Income per Capita                                     127
27. Extreme Poverty                                                131

Exacerbating Conditions and Events                                135
28. Dependency Ratio                                               137
29. HIV / AIDS                                                     140
30. Malaria                                                        143
31. Health Expenditures per Capita                                145
32. Unsafe Water and Poor Sanitation                              148
33. International Trade and Primary Products                       154
34. International Terms of Trade                                   157
35. Terms-of-Trade Shocks                                         160
36. Food Trade                                                    162
37. Food Price Shocks                                             165
38. Development Aid and Food Aid                                  168
39. War                                                            174
40. Natural Disasters                                             178

Conclusion                                                        181

Appendix 1. Map Data Sources                                      185
Appendix 2. Hunger Vulnerability Index                            188
Notes                                                             191
References                                                        193
Index                                                            199

# Maps

1.1 Safe drinking water — 8

1.2 People living on less than $1.25 per day — 8

2.1 Food supply as a percentage of daily need (2300 calories per day) — 15

2.2 Change in food availability — 16

2.3 Food supply as a percentage of daily need (2100 calories per day) — 18

3.1 Prevalence of undernourishment — 19

3.2 Change in share of population that is under-nourished — 22

3.3 Estimated number of undernourished people — 23

3.4 Change in number of undernourished people — 23

4.1 Vitamin A deficiency in preschool-aged children — 27

4.2 Iodine deficiency — 30

4.3 Prevalence of anemia — 31

5.1 Overweight and obese children 0–5 years of age — 34

5.2 Obese adults in the world — 35

6.1 Children with moderate or severe growth failure — 39

6.2 Change in child growth failure rate — 39

6.3 Change in the number of children with growth failure — 40

7.1 Population living on less than $2.00 per day — 48

7.2 Number of people living on less than $2.00 a day — 48

8.1 The Hunger Vulnerability Index — 56

9.1 Child growth failure in Uruguay by department — 59

9.2 Rural-urban differences in child growth failure in Uruguay — 60

9.3 Child growth failure in Montevideo by neighborhood — 61

10.1 Food insecurity in the United States — 63

10.2 Changes in hunger prevalence in US households — 64

10.3 Poverty per capita in the United States — 64

10.4 Percentage of poor households participating in Food Stamp Program — 65

11.1 Severe anemia in India — 67

11.2 Child growth failure in India — 68

11.3 Adult obesity in Mexico — 69

12.1 Population growth — 77

13.1 Arable land per person — 79

13.2 Change in arable land per person — 82

14.1 Environmental systems health — 84

15.1 Adult literacy — 86

16.1 Road density — 88

17.1 Change in resource base — 92

18.1 Projected impact of climate change on agricultural production — 94

19.1 College and university enrollment — 99

19.2 Research in the world — 100

19.3 Investment in research and development — 101

19.4 The digital divide — 101

20.1 Fertilizer consumption — 103

21.1 European colonies — 109

22.1 Foreign debt burden — 113

23.1 Political freedoms — 116

24.1 Income share held by poorest 10% — 118

24.2 Income share held by richest 10% — 119

25.1 Gender differences in life expectancy — 122

25.2 Gender-related Development Index — 123

26.1 Gross national income per capita — 128

27.1 Percentage of population in extreme poverty — 132

27.2 Change in rate of extreme poverty — 132

27.3 Population in extreme poverty — 133

27.4 Change in number of people in extreme poverty — 133

27.5 Number of people living in $1.25/day poverty — 134

28.1 Dependency ratio — 137

29.1 HIV infections among adults — 140

30.1 Total deaths from malaria — 143

31.1 Health expenditures per capita — 145

32.1 Improved drinking water — 148

32.2 Improved sanitation — 149

32.3 Cholera epidemic in Angola by province — 150

32.4 Indian households with improved drinking water — 152

| | | |
|---|---|---|
| 33.1 | Primary products as a share of merchandise exports | 154 |
| 33.2 | Trade (exports and imports) as a share of GDP | 156 |
| 34.1 | Average annual change in terms of trade | 157 |
| 35.1 | Largest single-year decline in terms of trade | 160 |
| 36.1 | Cereal trade | 162 |
| 37.1 | Food price inflation | 166 |
| 38.1 | Aid per capita | 168 |
| 38.2 | Food aid deliveries | 172 |
| 38.3 | Food aid recipients | 173 |
| 38.4 | Food aid per person living in extreme poverty | 173 |
| 39.1 | Military spending | 175 |
| 39.2 | People displaced by human-made disasters | 175 |
| 39.3 | Number of people affected by human-made disasters | 176 |
| 39.4 | Major armed conflicts | 176 |
| 40.1 | Number of people affected by natural disasters | 178 |
| 40.2 | Number of people affected by natural disasters, four-year annual average | 179 |
| 41.1 | Change in number of people undernourished | 183 |
| 41.2 | Change in proportion of population undernourished | 183 |

# Figures

1.1 Levels and measures of food (in)security 3

1.2 Sources of hunger 5

3.1 Undernourishment: change in number and change in prevalence 24

4.1 A fruit and vegetable seller and his assistants in Rajasthan, India 29

5.1 A family at a fast-food restaurant in Tamarindo, Costa Rica 35

6.1 WHO growth chart for boys ages birth to five years 38

6.2 Percentage of children experiencing growth failure 40

6.3 Number of children experiencing growth failure 41

6.4 Number of short-stature children 41

6.5 A group of children in Kericho, Kenya 41

7.1 A food vendor in Abidjan, Côte d'Ivoire 44

7.2 Percentage of population living on less than $1.25 per day 45

7.3 Distribution of $1.25/day poverty 46

7.4 Distribution of $2.00/day poverty 46

7.5 Percentage of population living on less than $2.00 a day 47

7.6 Distribution of population living on less than $2.00 a day 47

7.7 Poverty rate and child growth failure 49

7.8 Food availability and poverty 49

7.9 Poverty and growth failure in countries with adequate food availability 50

8.1 A Nepali cook making tea in the Shekawati area of Rajasthan, India 51

8.2 The shape and size of the hunger problem 52

9.1 Distribution of short-stature children in Uruguay 60

10.1 Lunchtime at a soup kitchen in east-central Illinois 66

12.1 Food production per capita in the world 76

12.2 Population growth and hunger vulnerability 77

12.3 Per capita carbon dioxide emissions for countries by economic group 78

13.1 Hunger vulnerability and arable land per capita 81

13.2 Mural showing farmers in Burkina Faso 83

14.1 Hunger vulnerability and health of environmental systems 85

15.1 Hunger vulnerability and literacy 87

16.1 Hunger vulnerability and road network 89

16.2 Hunger vulnerability and road network in countries with 25 to 250 people per square kilometer 89

17.1 "Bush fires? Never again!" Sign outside Pa, Burkina Faso 92

20.1 Farmer obtaining fertilizers on credit in northern Côte d'Ivoire 104

21.1 Portuguese government map showing the size of its colonies 110

21.2 Forced cotton deliveries in Tiébissou, Côte d'Ivoire, circa 1915 111

22.1 Debt service as a share of export revenue 113

22.2 Declining debt service under HIPC initiative 114

23.1 Political campaign in Korhogo, Côte d'Ivoire 115

23.2 Political freedoms and hunger vulnerability 117

24.1 Lorenz curve diagram 120

24.2 Hunger vulnerability and income inequality 120

24.3 Hunger vulnerability and income inequality in poor countries 121

24.4 Hunger vulnerability and income inequality in the poorest countries 121

25.1 Gender inequality and hunger vulnerability 123

25.2 Farmers in southern Burkina Faso 124

26.1 National income and hunger vulnerability 129

26.2 National income and undernutrition 130

26.3 Poverty rate and national income 130

28.1 Man with his two wives and nine children in Katiali, Côte d'Ivoire 138

29.1 HIV/AIDS education sign in rural Kenya 141

29.2 The projected demographic impact of HIV/AIDS in Botswana 142

31.1 Health spending and hunger vulnerability 146

31.2 Prices for maternity clinic services in Bougouni, Mali 147

32.1 Drawing water in rural Rajasthan, India 151

32.2 Percentage of population with access to improved water 152

33.1 Bales of cotton lint stacked outside a gin in Houndé, Burkina Faso 155

34.1 Francois Traoré at an APROCA meeting in Orodara, Burkina Faso 159

35.1 Terms-of-trade shocks and the prevalence of growth failure 161

36.1 Sacks of rice for sale in Abidjan, Côte d'Ivoire 164

37.1 Food price inflation and undernutrition 166

38.1 Elementary school students in Côte d'Ivoire 170

38.2 A bumper harvest in Champaign County, Illinois 171

39.1 United Nations peacekeeping troops patrol Korhogo, Côte d'Ivoire 177

# Tables

2.1  Change in number and percentage of
     countries within food supply classes by
     minimum caloric needs                          17
3.1  Undernourishment: Number and prevalence        24
4.1  Micronutrient deficiency rates by world
     region                                         26
5.1  Women's body mass index percentiles for
     selected countries                             32
5.2  Countries with the highest rates of
     overweight and obese children                  34
5.3  Percentage of adult women underweight          34
6.1  Countries with the highest levels of growth
     failure                                        37
6.2  Children under five years experiencing
     growth failure                                 40
7.1  National poverty lines                         43
8.1  The Hunger Vulnerability Index: Selected
     countries                                      52
8.2  Hunger Vulnerability Index: China              53
8.3  Contrasts between the Hunger Vulnerability
     Index and the Global Hunger Index              55
8.4  Contrasts between the Poverty and Hunger
     Index and the Hunger Vulnerability Index       55
10.1 Trends in food insecurity in the United States 63
10.2 Distribution of food insecurity in the
     United States                                  64
11.1 Trends in obesity among Mexican women          69

11.2 Rates of diabetes mellitus and hypertension
     among adults in Mexico                         69
20.1 Fertilizer use by region                      103
21.1 Top recipients of French and British official
     development assistance                        111
23.1 Selected freedom and hunger vulnerability
     ratings                                       117
26.1 Similar income per capita, contrasting
     hunger vulnerabilities                        129
27.1 Extreme poverty by region                     131
27.2 Share and number of people living in
     extreme poverty                               134
28.1 Dependency ratios and composition of
     dependents                                    138
28.2 Children orphaned by AIDS                     139
29.1 HIV/AIDS prevalence rates for the most
     seriously affected countries                  141
38.1 Food aid deliveries by region                 172
38.2 The top 10 recipients of food aid             172
40.1 Number of people affected by natural
     disasters                                     179
41.1 Change in number of undernourished
     people in the developing world                182
41.2 Change in prevalence of undernourishment
     in the developing world                       182
41.3 Correlation between hunger vulnerability
     and different variables                       184

# Boxes

| | | |
|---|---|---|
| 1.1 | Definitions of hunger and nutritional terms | 3 |
| 1.2 | How to read the maps in this atlas | 7 |
| 2.1 | Counting calories | 17 |
| 4.1 | Biofortification in practice: Orange sweet potatoes | 27 |
| 4.2 | Biofortification in practice: Golden rice | 28 |
| 5.1 | The nutrition transition and the double burden of malnutrition | 33 |
| 5.2 | Low BMI | 33 |
| 6.1 | Growth failure and growth charts | 38 |
| 7.1 | What does $2.00/day poverty mean? | 44 |
| 8.1 | Hunger vulnerability diamonds | 52 |
| 8.2 | Which index to measure hunger? | 54 |
| 12.1 | Population pressure and consumption pressure | 78 |
| 13.1 | The Landless Workers Movement of Brazil | 80 |
| 13.2 | When does arable land per capita relate to hunger vulnerability? | 81 |
| 20.1 | New Rice for Africa (NERICA) | 104 |
| 24.1 | Measuring inequality: Lorenz curves and Gini coefficients | 119 |
| 32.1 | Water privatization revolts in Bolivia | 153 |
| 33.1 | International trade, poverty and inequality | 155 |
| 34.1 | World cotton prices and hunger vulnerability in West and Central Africa | 158 |
| 35.1 | The coffee crisis | 161 |
| 36.1 | Biofuels, global food availability and prices | 163 |
| 37.1 | Food price dilemma | 167 |
| 38.1 | Food aid controversy in the United States | 169 |

# Acknowledgments

This atlas began as a class project in a development geography course taught by Bassett at the University of Illinois in the spring of 2003. The discussions that semester on the geography of hunger and its causes opened up new understandings and research directions that inspired this book. We are grateful to the students enrolled in that course for their participation in these discussions and for literally mapping the way forward in the student atlas produced that semester. Turning inspiration into a full-fledged book required a whole new level of commitment, so in early 2004, we decided over coffee in a local café to join our complementary skills and interests to write this atlas. The stimulating back-and-forth discussions of the maps and texts at our weekly meetings over coffee gave momentum to this collaborative undertaking.

The Rockefeller Foundation offered an extraordinary opportunity to reflect and work on this project at its Bellagio Center in Northern Italy. Although both of us were invited to this scholars' paradise in February 2006, family obligations prevented Winter-Nelson from traveling. With Bassett in Bellagio and Winter-Nelson in Urbana-Champaign, we revised the texts of 14 maps and drafted the introduction during this e-mail friendly residency. At the Bellagio Center, Pilar Palacia and her staff were exceptional hosts who made this work especially enjoyable for the half of the team in Italy. We are thankful to coresidents Teresa Porzecanski and Jaime Mendoza for leading us to the child growth failure data for Uruguay.

Over the course of writing and updating this book we were fortunate to have Moussa Koné and Treva Ellison serve as our research assistants. Their multiple contributions in data gathering, mapmaking, and reading the different map texts are apparent throughout this atlas. Collette DeJong, Jessica Palmer, Amy Podlasek, and Lenis Saweda Liverpool read the entire manuscript. We thank them and Karen Winter-Nelson for their many suggestions on how to make this work accessible and accurate. Carol Spindel's sharp blue pencil and incisive comments have helped to keep the prose clear and to the point.

The cartographic contributions of Ezekiel Kalipeni, Leo Zulu, Jane Domier, and Shannon Geegan have been indispensable at different stages of this project. We thank ColorBrewer.org for suggesting the map color schemes and Worldmapper.org for supplying the Java-based cartogram program. Kathleen O'Reilly's photographs of Rajasthan, India, expand the visual terrain of the following pages. All other photos are by Bassett except for figure 7.1, for which we thank Moussa Koné. Monika Blössner of the World Health Organization answered our many questions on the WHO's global database on child growth and malnutrition in timely and detailed e-mail messages from Geneva. Finally, we extend our gratitude to three anonymous reviewers and our editors, Christie Henry and Carol Saller, for their perceptive observations and suggestions on how to make this a better work.

To all, may your generosity be returned to you a thousandfold.

# 1: Introduction

*Things are good here now. The rich farmers eat twice a day and even the poor ones can eat once a day.*

**Small-scale farmer in Imdibir, Ethiopia**[1]

*We spend about $125–175 a month for groceries if not more. About the end of the month, things get real spare. You know, I got three teenagers, so about the end of the month I'm reducing to one meal a day, so I make sure the kids got everything they need.*

**Unemployed mother in Louisville, Kentucky, USA**[2]

## MAPPING HUNGER

In the United States, in Ethiopia, wherever there is poverty, there is also the likelihood of hunger. It is often a silent presence, but occasionally it roars. In Haiti, angry demonstrators gathered at the national palace in March 2008 to protest spiraling food prices. More than two-thirds of Haitians earn less than $2.00 a day. The poorest eat mud cakes made of dirt, oil, and butter. Their president, René Préval, dismissed their complaints, saying that if they could afford cell phones, they could afford to feed their families (Lacey 2008). Protestors responded to these callous remarks by banging on the palace gates and yelling that they were starving. In the ensuing riots, five people were killed while raiding a food warehouse. The Haitian senate ousted the prime minister and the government collapsed. Weeks later Préval announced a 15% price reduction for a sack of rice.

Hunger protests spread around the world that spring as prices for wheat, rice, and corn rose to twice their 2006 levels. Hunger vulnerability shook the gates of presidential palaces in dozens of countries, and governments scrambled to come up with emergency responses. For a moment, the world's attention focused on the food crisis, and many began to ask why there was so much hunger, and why it was so widespread. This atlas is a response to these questions and concerns. It focuses on the sources and geography of hunger, and provides insights for understanding and ultimately eliminating this unnecessary suffering in today's world.

Enough food is produced in the world for every person to enjoy a healthy diet. Yet about one billion of the world's six billion people eat too little to meet their basic caloric needs (USDA 2008, 3); two billion live on diets that are so deficient in specific vitamins or minerals that their health is at risk (WHO 2008h). About 30% of the children in developing countries have stunted physical development due to hunger, and each year eight million children under five die as a result of the interaction of poor nutrition and disease.[3] While millions suffer daily for lack of food, large quantities of grain are used to produce biofuels, animal feed, and sweeteners.

The coexistence of food abundance and widespread hunger presents a troubling conundrum. To unravel it, this atlas addresses two basic questions: "Where are the hungry?" and "Why are they hungry?" In answering these two questions, a third immediately arises: "Is hunger inevitable?"

There are both conceptual and methodological challenges to mapping the geography of hunger. How can hunger be measured? A number of indicators exist. How well do they actually describe and locate hunger? This atlas examines the strengths and weaknesses of the most commonly used indicators and then presents an alternative gauge, the Hunger Vulnerability Index (HVI). After mapping the distribution of hunger, the atlas explores its causes. A number of social, political, economic, and environmental themes are taken up to determine whether relationships exist between factors like gender inequality or population growth within a country and that country's vulnerability to hunger.

Food protests by hundreds of thousands of people in some 30 countries in 2008 showed that hunger vulnerability, the state of being at risk of hunger, is pervasive and demands urgent attention. Debates during the United Nations' 2008 summit on the food crisis brought out the complexity of the world hunger problem. International institutions like the World Bank and the United Nations Food and Agriculture Organization (FAO) pushed for increasing funding to low-income countries for agricultural development and food aid and for reforming trade policies that contribute to food price inflation (*BBC News,* June 5, 2008). The nongovernmental organization Oxfam America declared that agricultural subsidies and biofuel policies in rich countries contribute to world hunger and need reform (Oxfam 2008). Civil society groups like the smallholder farmer organization Via Campesina explained the global food crisis as the outcome of decades of trade "liberalization" policies that permit rich countries to dump their agricultural surpluses onto markets and thus undermine local peasant and family farm production (Via Campesina 2008). And many people maintained the idea that population growth is the main cause of hunger—that there are simply too many people making too many demands on food-producing resources. The complexity of the problem seems daunting.

The roots of hunger are tangled and complex, but hunger is always closely tied to poverty and social vulnerability. The physical environment, social norms, technology, and market conditions all factor into hunger vulnerability. However, focusing on these issues alone can hide the role of governments and political economy in diminishing or exacerbating their impact. This atlas treats government and political economy as central and sees hunger as ultimately rooted in policies that leave people too poor to cope with their social and environmental conditions. The maps and accompanying texts, graphs, and photographs build upon this conceptual approach.

The first purpose of this atlas is to show the geography of hunger in order to discuss conditions that place people at risk. The second objective is to contribute to international efforts to reduce world hunger. At the 1996 World Food Summit, the heads of state of 186 countries adopted a resolution to reduce by half the absolute number of undernourished people in the world by the year 2015. This goal was subsequently taken up in modified form by the United Nations in its Millennium Development Goals (MDG). Goal number one of the MDG is to eradicate poverty and hunger by cutting in half the percentage of people in the world

who are suffering hunger and living on less than $1.00 a day. This atlas provides information on the human geography of hunger to support people and institutions developing food security policies and programs.

The third objective is to provide educational materials to teachers, students, and anyone else who might use this atlas to inform opinions and actions concerning nutrition, population and health, human rights, poverty, and international development. Hunger is a complex phenomenon that reveals much about the societies and places in which it occurs. The following maps reveal that there is a geography of hunger. It is often located in countries where social inequalities are high and poverty is widespread. We hope that this work will deepen each reader's understanding of hunger around the world and thus map a way toward changing the conditions that produce deprivation in the midst of plenty.

## MEASURES AND LEVELS OF FOOD (IN)SECURITY

Hunger can be analyzed using a range of measures and geographic levels; depending on which indicators and geographic scales we choose, different aspects of the hunger problem come into focus. Food deprivation can be described in many ways, so the terms used in the atlas are defined in box 1.1. Of these terms, *hunger* (the physical pain from eating too little) is often the most relevant to those suffering, but the most difficult to measure. Figure 1.1 presents three important measures and levels that are central to our analysis of hunger. At a high level of aggregation, we examine national food availability. This measure estimates the availability of calories in a country and divides that total by the country's population. Comparing per capita caloric availability with a minimum dietary requirement can indicate whether a country has the capacity to feed itself through production and imports. The FAO annually produces national food balance sheets that provide data on per capita caloric availability for most countries (map 2.1).

Countries whose national food availability level is close to or below the minimum requirement are bound to have widespread malnutrition. But hunger is often common where food availability is adequate because food is unevenly distributed across households. The

**Box 1.1. Definitions of Hunger and Nutritional Terms**

*Hunger:* The physical pain and discomfort associated with the inadequate consumption of food. Hunger is an outcome of food insecurity manifested in undernourishment and undernutrition.

*Hunger vulnerability:* The likelihood of current or future exposure of an individual or group to hunger. Hunger vulnerability is linked to socioeconomic, political, and biophysical processes operating at local, national, and international scales.

*Food Security:* Reliable access to sufficient quantities and types of food to ensure an active and healthy life. There are four basic elements to food security: food availability, access to food of sufficient quality and quantity, a stable supply, and a culturally acceptable food.

*Food Insecurity:* The absence of food security associated with inadequate or unstable supplies of or insufficient access to culturally acceptable food.

*Undernourishment:* Inadequate food intake to meet caloric requirements.

*Malnutrition:* Poor nutritional health due to inadequate or excessive absorption of protein, energy, and/or micronutrients in a diet. Malnutrition is commonly measured in children below the age of five by wasting (low weight for height), growth failure (low height for age), and other clinical signs.

*Undernutrition:* Malnutrition due to insufficient nutrients in a diet or poor absorption of nutrients in the diet.

*Overnutrition:* Malnutrition due to excessive consumption relative to physical activity. Overnutrition is manifested in obesity.

*Micronutrient deficiency:* Undernutrition due to the inadequate intake or absorption of essential vitamins and minerals especially Vitamin A, iron, zinc, and iodine. These micronutrients are vital to the production of enzymes and hormones that are necessary for growth, reproduction, and immune system development. They are particularly important during early childhood, pregnancy, and other periods of rapid growth.

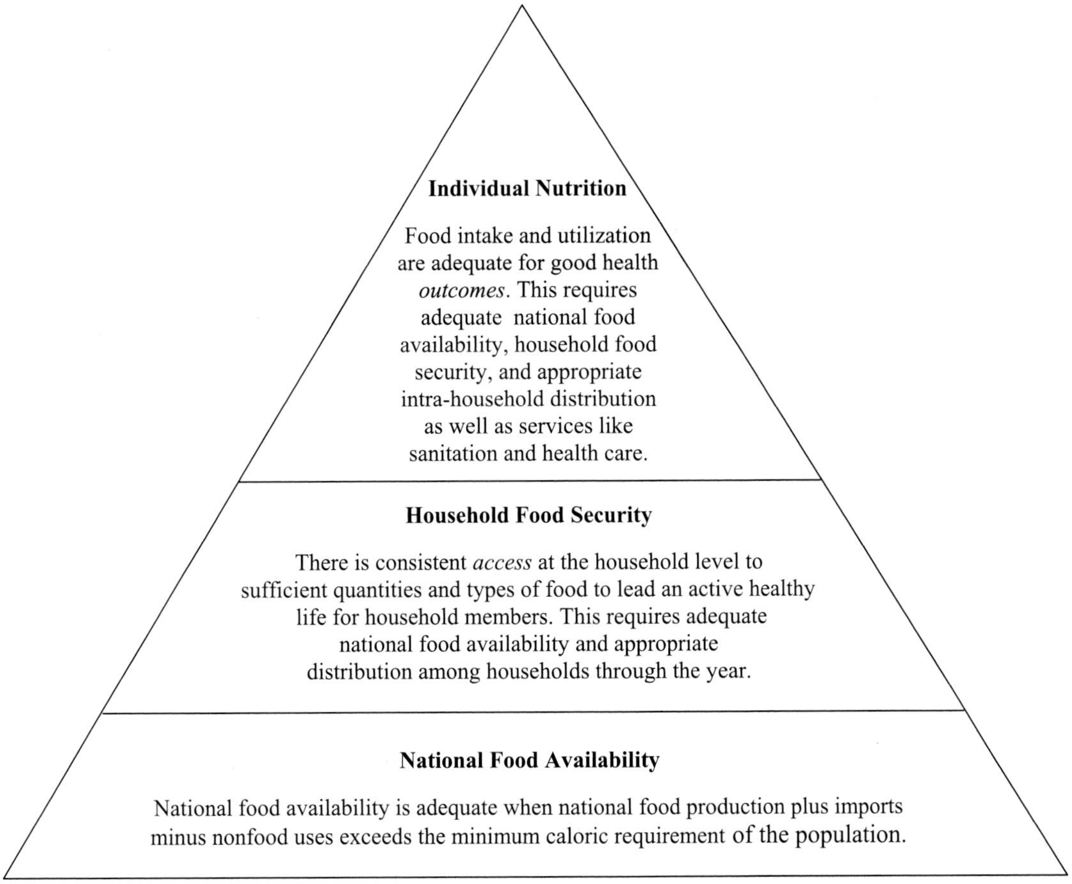

**Individual Nutrition**

Food intake and utilization are adequate for good health *outcomes*. This requires adequate national food availability, household food security, and appropriate intra-household distribution as well as services like sanitation and health care.

**Household Food Security**

There is consistent *access* at the household level to sufficient quantities and types of food to lead an active healthy life for household members. This requires adequate national food availability and appropriate distribution among households through the year.

**National Food Availability**

National food availability is adequate when national food production plus imports minus nonfood uses exceeds the minimum caloric requirement of the population.

Figure 1.1. Levels and measures of food (in)security.

notion of household food security moves from availability of food at the national level to access at the household level. While sufficient national food availability does not guarantee household food security, food security cannot exist for all households unless national food sufficiency is met. Thus, household food security is built on national food availability in figure 1.1. The FAO measure of undernourishment (map 3.1) combines data on national food availability with information on food distribution across households to derive a measure of household food security.

Since household food security influences individual access to food, individual nutrition rests on household food security, as reflected in figure 1.1. Reliable access to sufficient food at the household level insures adequate nutrition only if other factors are in place. To move from a measure of household food security to one that captures the nutritional status of individuals requires collecting data directly from the people in question. These data can come from studies that measure physical features like the prevalence of growth failure or vitamin A deficiency. In the context of well-constructed studies, these sample surveys can be aggregated to yield estimates of the intensity of specific nutritional problems in a population. Unfortunately, such studies are not uniformly conducted across regions. The World Health Organization (WHO) of the United Nations has constructed a large pool of information on children's weights, heights, and ages, but most other specific indicators of nutritional outcomes, such as wasting among adults or micronutrient deficiencies, are not so well tracked.

There is a value to measuring food issues at each of these scales: national food availability, household food security, and individual nutritional outcomes. Doing so helps identify appropriate intervention points by providing as complete a picture of the food situation as possible.

## THE SOURCES OF HUNGER

The food price protests of 2008 revealed that hunger is fundamentally bound to poverty and social vulnerability. Indeed, hunger may be the most obvious manifestation of poverty; and as long as the causes of poverty remain, hunger will persist. Figure 1.2 presents poverty as rooted in the interplay of resources, technolo-

gies, and power relations that emerge through social institutions.[4]

All societies are endowed with a set of resources. A society's resource base includes its natural resources such as forests, minerals, water, and soils, and also its built capital such as transportation infrastructure and utility grids. Human assets constitute another class of resources and can be thought of as the productive capacity contained in individuals, which is enhanced by their education and health.

Technology sets the limit on what and how much can be produced from a given resource base. Development of new technologies can enable a society to produce more food or other goods from its resources. Changes in resources and technologies are related. High stocks of human assets, for example, can promote the development of new technologies which themselves might conserve (or exhaust) elements of the natural resource base.

Every society has a set of institutions, or formal and informal rules, that govern access to resources and use of technology. Important institutions include those that influence political processes and set terms for market exchanges. Examples include electoral procedures, regulations on international trade, food price controls, and land tenure systems. These rules often reflect power relations in a society and they likewise influence the distribution of power over resource allocation. While resources and technologies set limits on what can be produced, institutions and power relations determine what is actually produced and who gets it.

In any society, political and market institutions determine the public and private goods and services that emerge from the available resources and technologies. Public goods can include such things as the systems of health care, education, sanitation, and law enforcement that are shared by the citizenry. Private goods and services are those that can be owned by individuals, like clothing, corn, cars, and haircuts. It is the volume, form, and distribution of public and private goods that determine the incidence of poverty. Poverty may arise because too little is produced, because access to public goods and services is inadequate, or because the production of an economy is distributed in a highly uneven manner. In any case, poverty and hunger are results of the interplay of resources, technologies, and

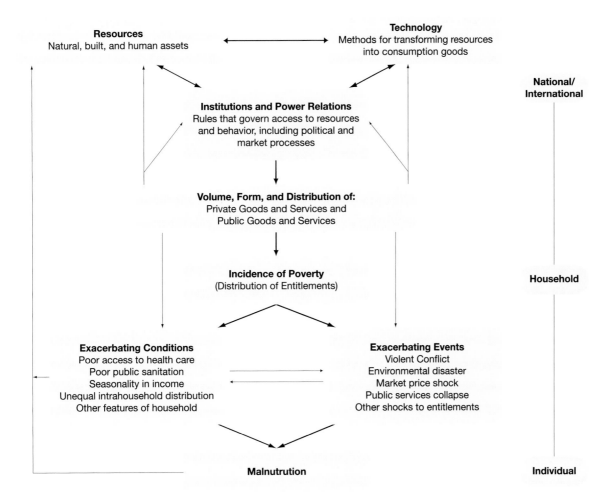

Figure 1.2. Sources of hunger.

power relations as they affect people's access to goods and services.

Poverty and social vulnerability can be related to hunger through the notion of entitlements. Amartya Sen, who received the Noble Prize in economics in 1998, defined a person's entitlement as the various combinations of things she could acquire through legal channels. Hunger is inevitable when the distribution of goods and services leaves some households with entitlements that do not include enough food for all of their members. The entitlement depends on the availability of public and private goods and services. A household could be exceedingly poor in terms of private income, but if public services ensure that food is available to the poor, the household's entitlement would include sufficient food to avoid nutritional prob-

lems. Brazil's Zero Hunger Project is an example of a public program designed to bring food security to poor households. Food stamps in the United States represent another effort to avert hunger by expanding the entitlements of the poor.

Poverty is usually defined at the level of the household. A household is in poverty when it is not certain that its entitlement will provide for its basic needs such as food and shelter. Hunger, in contrast, is suffered by individuals within households. The presence of certain exacerbating conditions and events can push people in impoverished households into hunger.

Around the world, many people live under conditions that make their poverty more likely to result in malnutrition. These conditions include unequal distribution within the household which can leave some members disadvantaged, and poor sanitation systems that make diarrheal diseases inevitable and result in poor nutritional outcomes despite otherwise sufficient food intake. The list of conditions that exacerbate poverty to make hunger more prevalent is long. A few examples are included in figure 1.2.

Aside from persistent conditions that intensify the effect of poverty on hunger, the poor are additionally susceptible to hunger as a result of catastrophic events. These include violent conflict (war in Sudan), environmental disaster (tsunami in Southeast Asia) and market shocks (rising food prices around the world). Such cataclysmic events can diminish a household's entitlement, exacerbate its poverty, and cause some or all of its members to go hungry. When the event affects a large share of a population and governments fail to restore entitlements through relief efforts, the result is famine.

The elements pictured in figure 1.2 are not static. Over time there are important feedbacks among them, indicated by the thin lines in the figure. For example, while resources help define the potential for production and food consumption, the incidence of malnutrition also has a direct effect on the resource base. Nutritional deficits affect mental and physical development, morbidity, and mortality (human assets); and insufficient access to food leads some households to consume natural resources (seeds, livestock). Many of the exacerbating conditions that make poverty more likely to result in hunger also affect the resource base and ultimately the prevalence of poverty. Women's limited access to education, for example, can intensify the effect of poverty on hunger because uneducated women cannot find good paying jobs and thus cannot buy food when needed. Women's low education levels also undermine the stock of human assets and affect future poverty, even if hunger does not emerge in the present.

There is also a feedback from public and private goods to resources, technologies and institutions. A larger supply of public education and research, for example, can spur the development of new technologies; the distribution of private goods and services can result in pressures on institutions and changes in power relations; and the mix of public versus private investment can determine the nature of infrastructure development and people's health and education.

Figure 1.2 also indicates the different scales at which hunger problems are rooted and experienced. Malnutrition and hunger are experienced by individuals, while the poverty that allows hunger is a feature of households. The technology, resources and power relations that influence poverty are national in scope.

The conceptual framework illustrated in figure 1.2 greatly simplifies the causes of poverty and hunger around the world. It does, however, direct attention to the relationships among key variables and the incidence of poverty. The consequences of poverty are often made worse by humanitarian crises linked to violent conflicts, droughts, floods, and other disasters. News reports typically emphasize these exacerbating conditions and catastrophic events in their coverage of food crises. Hunger is often associated with these factors, but it is ultimately linked to the larger political and economic forces that make people vulnerable.

There is a large social science literature that examines the political and economic dimensions of hunger vulnerability (Scott 1976; Sen 1981; Watts 1983; United Nations 1997). These studies commonly distinguish three types of causes in their explanations of hunger. Immediate causes of hunger are poor food consumption and disease. Underlying causes of malnutrition and disease include poverty, inadequate health services and sanitation, and poor child feeding and care practices. Ultimate causes encompass the broader social and political-economic context and include power and class relations, economic distribution, ideologies, resources, and technology. In our framework (figure 1.2), we emphasize poverty and entitlement failure as the most important underlying causes of hunger. Poverty and entitlement failure are linked to institutions and power relations through the notion of empowerment, which refers to the capacity of individuals and groups to shape the political economy that determines entitlements (Watts and Bohle, 1993; Ribot 1996). Hunger vulnerability thus measures the susceptibility of people to food shortages in their particular political economies. Vulnerability will be high where governments are unaccountable to their people and entitlements are weak. It will be lower where capable and responsive government is combined with strong entitlements.

In February 2007, tens of thousands of people marched through the streets of Mexico City protesting the high price of tortillas. The price of this Mexican

**Box 1.2. How to Read the Maps in This Atlas**

All maps reflect the objectives of their makers. Our objectives in this atlas are to show where hunger exists in the world and to explain its geography. Each map in this atlas has four elements: title, map, legend, and source. The title informs the reader of the map's theme or subject; the map shows the distribution of that subject based on different data groups and colors; and the legend provides information on the data groups. The full citation for each data source is listed at the end of this atlas. The text that accompanies each thematic map discusses the nature and quality of the data, connects the subject to the larger themes of the atlas, and comments on the geographical patterns that appear in the map.

The map of safe drinking water in the world illustrates these map components (map 1.1). The title tells the reader the map's theme: safe drinking water. The legend indicates that the data displayed in the map refer to the percentage of the population in each country (for which we have data) that drinks from an improved water source. Exactly what is meant by an improved water source is discussed in the accompanying text. The legend breaks down into five color classes to help the reader interpret the map. For example, the lightest yellow and green colors refer to those countries in which less than 50% of the population has access to safe drinking water. Two countries that fall within this range are Afghanistan (22%) and Chad (48%). The breaks are defined so that all data points that are below the minimum of one group are included in the next lower group.

The map itself shows the distribution of safe drinking water across the world. A glance at the map indicates that a large percentage of people in African countries lack access to improved water supplies, while North Americans are much better served. At closer examination, a diversity of situations appears within Africa and Southeast Asia.

Note that data are mapped at the country level. Information is not displayed at the geographical scale of the state, province, or department of the countries. This is because data are rarely available at the subnational scale. Data are most often compiled at the national level by international institutions like the World Bank and United Nations. These institutions and their publications are the primary sources of information for this atlas. That said,

some maps in this atlas present data at the subnational level to show how the distribution of a particular phenomenon varies within a country. The examples of child growth failure in Uruguay and hunger in the United States reveal how hunger varies within a country (maps 9.1–10.3). Such examples show that hunger is not found everywhere in a country. Rather, it tends to be concentrated in certain regions and localities. Our goal in making such maps is to demonstrate this uneven geography of hunger. We also hope to encourage the atlas user to reflect on the origins of these geographical patterns.

The color scheme of the maps and legend typically goes from lightest to darkest with the darker shades indicating higher incidence of the phenomena being mapped. In the case of safe drinking water, the countries colored in the darkest shade of blue have the highest clean water coverage; those colored in the lightest shades are the least served. Countries for which no data are available are left white in these maps.

Finally, the sources of map 1.1 are the World Health Organization and the United Nations Children's Fund (WHO-UNICEF) and the World Bank. These organizations collect information and monitor changes in improved water supply and sanitation in the world. The Internet site that provides data on these topics is listed in appendix 1 at the end of the atlas. Atlas readers can use this source list to learn more about a particular region or country's water and sanitation situations.

We use the Eckert IV map projection throughout the atlas. This is an equal area projection that shows the actual size of a country or continent with relatively little distortion of its shape. A few of the maps in this atlas, however, greatly distort the size and shape of the earth's surface. These are cartograms, maps that modify the size and shape of country in relation to the variable being mapped. Map 1.2 is an example of this second type of map projection. It shows the distribution of people living on less than $1.25 a day. North America is exceedingly thin while Asia is extraordinarily large. Each country's distorted shape is proportional to the number of people residing there who live on less than $1.25 a day. Even without reading the legend, the map reader can see that most of this poverty is located in Sub-Saharan Africa and especially South and East Asia.

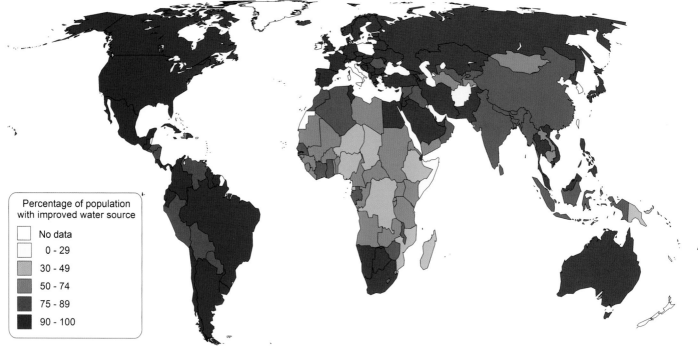

Map 1.1. Safe drinking water, 2002–6.

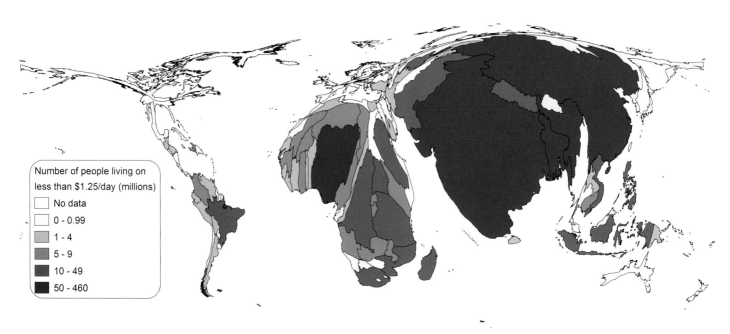

Map 1.2. People living on less than $1.25/day.

staple had increased 40% in three months, forcing poor households to spend a large share of their daily incomes on food. The price increases were partly a result of greater demand for corn in the United States by the ethanol industry, which pushed corn prices to their highest levels in ten years (McKinley 2007). The imme-diate cause of this crisis in Mexico was food becom-ing too expensive to allow poor households sufficient consumption. An underlying cause was poverty. More than one out of six Mexicans live below the national poverty line, earning too little to meet minimum needs of food, clothing, and shelter.[5] This widespread poverty left many people vulnerable to hunger when corn prices rose. An ultimate cause of hunger vulnerability in Mexico is the skewed income distribution that gives the richest 10% of the population 40% of the national income while leaving the poorest 10% with about 1.5% of the total income. The popular resistance to the market shock garnered an immediate concession from president Felipe Calderón, who reversed his usual advocacy of free markets and set price limits on corn flour. The empowered protestors' demands went further, calling for a renegotiation of the North American Free Trade Agreement to protect Mexican farmers and consumers from price movements abroad. Under the nationalist banner of "food sovereignty," demonstrators aimed to restructure the Mexican political economy. The protestors' demands pointed to the sources of hunger and to policies that could reduce their heightened vulnerability. The nature of these demands and apparent concessions illustrate the importance of politics and ideologies in shaping political economies, entitlements, and vulnerabilities.

# Part I: Locating Hunger

Our ability to locate hunger depends on the availability and quality of data that demonstrate its existence. In the first part of the atlas we assess the strengths and weaknesses of the most common indicators used to map world hunger. We then present an alternative measure, the Hunger Vulnerability Index (HVI), which builds upon the strengths and minimizes the weaknesses of prevailing indicators. The map of hunger vulnerability (map 8.1) tells us where people are vulnerable to hunger at the level of each country. It does not, however, reveal anything about the geography of hunger within a country. In the final section of part I, we examine the distribution of hunger at the subnational level to demonstrate the advantages of collecting data and mapping hunger at a variety of scales. Together the maps reveal the distribution of hunger among and within countries, which is important for policy makers who need to target interventions to the most vulnerable places and populations.

The maps assembled in part I also show that a single map can oversimplify patterns of hunger. Two maps are commonly made to present the FAO's prevalence-of-undernourishment indicator. The first shows the *percentage* of undernourished people by country, while the second displays the *number* of undernourished people in the same countries. These two maps present strikingly different images of world hunger. They show that the cartographic challenge of locating hunger is inextricably tied to the indicators we use to construct our maps.

# Indicators of Malnutrition

What are the conventional indicators of hunger? How good are they in locating and tracking hunger over time? In this section we first examine two of the most common measures of hunger: food availability, which indicates the number of calories available per person in a country, and the prevalence of undernourishment (POU), the FAO's indicator that considers both availability and distribution. The POU measure is the indicator most frequently used by UN organizations to track progress in reaching the Millennium Development Goal of cutting in half the prevalence of hunger in the world by 2015. We then address the quality of diet, particularly the debilitating diseases that result from insufficient amounts of vitamin A, iron, and iodine. We also consider the problem of obesity, a form of malnutrition linked to the excessive consumption of foods that are high in fats, starches, and carbohydrates.

The remainder of the section examines alternative indicators of hunger at the household and individual levels. The percentage of short-stature children in a country indicates chronic malnutrition. We select this variable as one component of the Hunger Vulnerability Index because it is an unequivocal measure of undernutrition at the individual level. The percentage of a country's population that earns less than $2.00 a day is a good indicator of households with unreliable access to food. We select this variable as our second component in the HVI. We then construct the HVI as our alternative measure of hunger. The index is comprised of three components (food availability, $2.00/day poverty, and child growth failure) which correspond to access to food at the national, household, and individual levels. In part II of the atlas, we use the HVI in our assessments of the potential causes of hunger.

# 2: Food Availability

How do we know if people are consuming sufficient amounts of food in a country? If we knew the answer to this question, we would have a better grip on the geography of world hunger. But it would be extremely time-consuming and expensive to measure the actual quantity and quality of food that every individual eats. When first confronted with this problem, the FAO approached the question from a different angle. It asked "How do we know if sufficient food exists in a country to feed its population?" The organization developed a method known as the food balance sheet that provides data on a country's food supplies and uses. The food balance sheet estimates the amount of food available for human consumption as: total food production and imports minus exports, feed, seed, industrial uses, changes in national reserves, and waste. The FAO uses the amount of food available to people to indicate the food consumption levels of a population as a whole (FAO 1998). Map 2.1 shows the geography of food avail-

ability according to the FAO's food balance sheets. By comparing food availability per person to an estimated food requirement, the map indicates countries where supplies of food are adequate or deficient. The significance of this map depends on how well food availability is measured, and what availability means for hunger.

The food balance sheet method relies on government agencies in each country to collect data on the production, trade, and use of agricultural products. These agencies must first determine the amount of food produced, imported, and exported, as well as that portion held in reserves. One problem with this initial calculation is that the national production figures typically underestimate noncommercial food production. That is, significant amounts of food grown and hunted for home consumption may be excluded from the "food produced" category. One study estimates underreporting to be as high as 25% (Svedberg 1999). In addition, the data on food exports and imports assumes

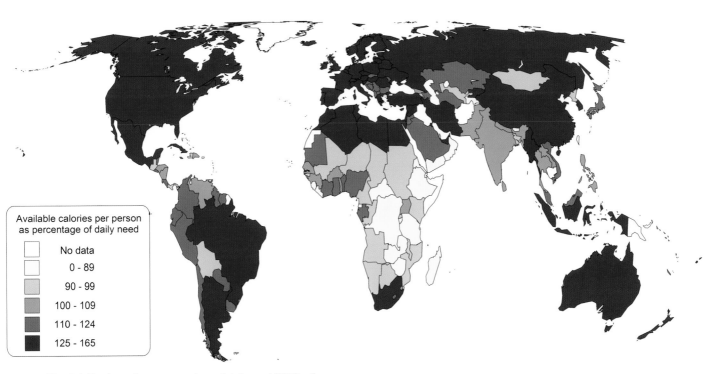

Available calories per person as percentage of daily need

No data
0 - 89
90 - 99
100 - 109
110 - 124
125 - 165

Map 2.1. Food supply as a percentage of daily need (2300 cal).

that a government accurately records all trade crossing its international borders. FAO acknowledges that in some countries smuggling foodstuffs is common and goes unrecorded.

The FAO then calculates the amount of food that is consumed by animals, processed into nonfood products like paper or ethanol, used for seed, or simply wasted. Once the total supply of food available for human consumption is known, its caloric value is estimated by FAO. The total caloric value of a country's food supply is then divided by its population and then by 365 days to indicate the number of calories available per person per day.

For example, the FAO's 2005 food balance sheet for Brazil shows that there were 3270 calories available per person per day that year (FAOSTAT data, in FAO 2007a). Due to losses from cooking and wastage, and some unevenness in distribution, actual consumption in the median household will be below this availability level. Nutritionists consider availability of 2100 to 2300 calories per day to be necessary for people to lead healthy and active lives.[6] Based on the 2300-calorie threshold, Brazil has more then enough food available to feed its population. Indeed, the map shows Brazilians to enjoy a level of food abundance that approaches that of the United States and Canada. With the exception of Bo-

livia, where just 97% of the population's food requirement is available, and French Guiana, for which data are not available, most South American countries possess adequate food supplies at the national scale. We know, however, that hunger is widespread in many of these countries. There are millions of hungry people in Brazil. Twenty-two percent of the children in China suffer growth failure due to poor nutrition (WHO 2008c). Yet, as the map shows, China has more than enough food for its population of 1.3 billion. It would be a mistake to treat a map of food availability as a map of hunger.

Why does this map fail to capture these nutritional realities? The food balance sheets inform us about available food supplies at the national scale but tell us little about actual food consumption at the level of the household and individual. When one equates national food availability with food security, as frequently occurs (Allen 1998, 90), these important distributional distinctions are lost. We know, for example, that poverty and sharp inequalities in income prevent millions from producing and/or buying sufficient food. Governments also fail to help poor households meet their basic food needs through entitlement programs. As a result, household scarcity coexists with national abundance.

What does the food availability map tell us? It points to the strong likelihood of hunger in Guatemala, Pan-

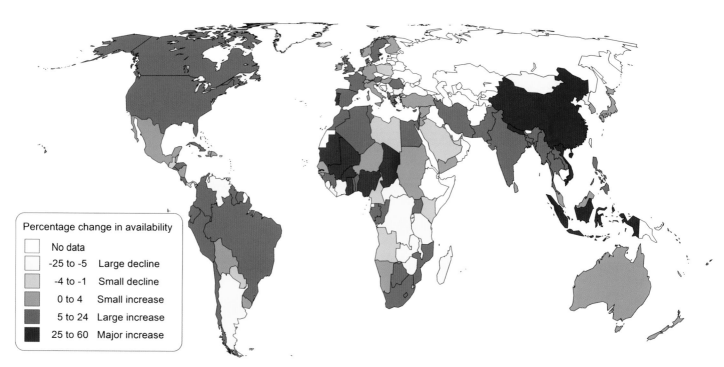

Percentage change in availability

| | |
|---|---|
| No data | |
| -25 to -5 | Large decline |
| -4 to -1 | Small decline |
| 0 to 4 | Small increase |
| 5 to 24 | Large increase |
| 25 to 60 | Major increase |

Map 2.2. Change in food availability, 1979–81 and 2001–3.

**Box 2.1. Counting Calories**

Opinions vary about the number of calories a person needs to consume to lead a healthy and active life. The United States Department of Agriculture (USDA) sets 2000 calories per day as a healthy diet, while the United Nations Food and Agriculture Organization (FAO) considers about 1800 calories to be sufficient for most people. Below are examples of 1800- and higher-calorie diets.

*Breakfast*

| | |
|---|---|
| ½ cup cornmeal (for porridge) | 200 calories |
| 1 tablespoon margarine | 100 calories |
| 1 cup whole milk | 150 calories |
| Tea with 3 teaspoons sugar | 50 calories |
| SUBTOTAL | 500 calories |

*Lunch*

| | |
|---|---|
| 1 cup rice | 250 calories |
| 2 eggs cooked in | 200 calories |
| ½ tablespoon oil | 50 calories |
| SUBTOTAL | 500 calories |

*Dinner*

| | |
|---|---|
| Tea with milk and sugar | 200 calories |
| 1 cup cooked rice | 250 calories |
| ¼ cup lentils | 175 calories |
| Onion, carrot, chicken broth | 75 calories |
| 1 tablespoon oil | 100 calories |
| SUBTOTAL | 800 calories |

| | |
|---|---|
| TOTAL | 1800 calories |

To make a 1900-calorie diet, add a banana, a half a doughnut, or just under a half cup of rice. To make a 2100-calorie diet, add all three of the above items, or two 12-ounce glasses of a soft drink. By comparison, you could go to McDonalds and have an Egg McMuffin with an order of hash browns for breakfast (430 calories), a Hot 'n Spicy Chicken Sandwich with a large soft drink for lunch (750 calories), and a double cheeseburger for dinner (760 calories). That would be a 1940-calorie day. Add an order of fries (540 calories) and a large soft drink (310 calories) to your dinner for a 2790-calorie day.

ama, Haiti, and Bolivia, in much of Central and East Africa, and in the Asian countries of Bangladesh, Tajikistan, and Mongolia. These countries and regions simply do not have enough food available at the national level to meet nutritional requirements. The map further suggests that modest distributional inequalities between or within households will produce hunger in countries like Venezuela, Pakistan, and India. Map 2.2, showing changes in availability over time, reveals regions where food availability is declining. Such areas, like the Democratic Republic of the Congo in Central Africa, are likely to see increasing food prices and growing hunger. For countries like Brazil and the United States where food supplies are abundant and growing at the national scale, widespread hunger can emerge only under gross inequality in distribution.

Twenty-three hundred calories per day is a common threshold set by scholars and agencies to indicate adequate food availability for a country's population. But some consider 2100 calories to be adequate. The lower level may be justified if there is less wastage or less inequality in distribution so that people's consumption is closer to availability, or if one thinks the caloric

requirement for healthy living is lower. For example, the USDA sets the appropriate daily consumption level for the US population at 2000 calories, but the FAO considers 1800 calories appropriate for many populations. Box 2.1 gives some idea of what these quantities of calories represent in terms of actual foods.

How would a geography of hunger change if the availability of 2100 calories per day were considered sufficient? Map 2.3 and table 2.1 reveal significant changes

**Table 2.1. Change in number and percentage of countries within food supply classes by minimum caloric needs**

| Availability | 2300 Calories per day | | 2100 Calories per day | |
|---|---|---|---|---|
| | # of Countries | % of Countries | # of Countries | % of Countries |
| Grossly inadequate | 14 | 10 | 5 | 3 |
| Inadequate | 26 | 18 | 14 | 10 |
| Barely adequate | 16 | 11 | 21 | 14 |
| Adequate | 31 | 21 | 23 | 16 |
| Abundant | 58 | 40 | 82 | 57 |
| Total | 145 | 100 | 145 | 100 |

Note: Grossly inadequate = 0%–89% of caloric needs. Inadequate = 90%–99% of needs. Barely adequate = 100%–109% of needs. Adequate = 110%–125% of needs.

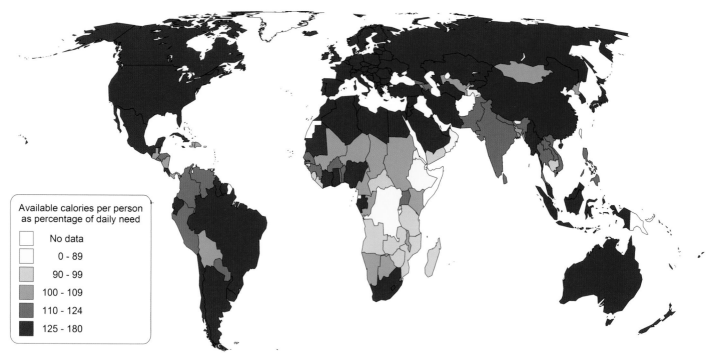

Map 2.3. Food supply as a percentage of daily need (2100 cal).

in the number of countries in each of the food supply categories that would result from lowering the caloric threshold. The general impression is that hunger is not as severe a problem at the global scale. When 2300 calories per day is used as the threshold, 28% of the countries are classified as having grossly inadequate or inadequate food supplies (table 2.1). This percentage drops to just 13% of all countries at the 2100 calories per day cutoff point. A comparison of maps 2.1 and 2.3 demonstrates that the threshold one uses to define inadequate food supplies can make an enormous difference in how the geography of hunger is mapped.

Food availability is an important part of the hunger problem. When food becomes less available for human consumption, prices rise and the poor are at increased risk of hunger. But food availability does not ensure access to food by those in need. In the end, the food availability map does not tell us where hunger exists because it does not capture the distributional dimension of food consumption at the household and individual levels. We need better measures and maps that show the presence of hunger in the world.

# 3: Prevalence of Undernourishment (POU)

The FAO is well aware of the limitations of food balance sheets as indicators of hunger in the world. It has long recognized that food availability (map 2.1) does not reveal much about differences in diet among different people and regions in a country.[7] In an attempt to better measure the frequency of hunger in national populations, the FAO constructed an indicator called the *prevalence of undernourishment* (POU). This measure is now used by many international agencies to track progress toward alleviating world hunger. The POU is calculated from three components: (1) an estimate of per capita food availability in a country; (2) an estimate of the distribution of available calories among households in the country; and (3) an estimate of the minimum daily caloric needs and caloric cutoff points below which individuals could be labeled "undernourished." With these data, the prevalence of undernourishment can be calculated based on the share of the population for which caloric availability at the household level is below the cutoff point.

Map 3.1 presents a geography of hunger based on the measured prevalence of undernourishment. As a representation of hunger, this map is clearly an improvement over map 2.1. It shows for example that hunger is present in countries like Brazil and China where we know it exists despite national food sufficiency (map 2.1). Still, the undernourishment measure is an imperfect gauge of hunger in the world. There are multiple concerns.

First, to obtain information on the availability of food in each country, the FAO relies on its food balance sheet data. As discussed in the text relating to map 2.1, these data are not reliable when a large share of food is produced and consumed by small-scale producers, as is the case in much of Africa and Asia. Errors in estimating production are compounded by poor

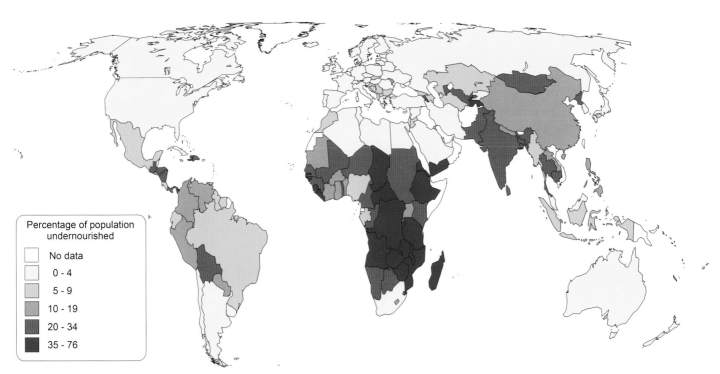

Percentage of population undernourished

No data
0 - 4
5 - 9
10 - 19
20 - 34
35 - 76

Map 3.1. Prevalence of undernourishment, 2002–4.

information about postharvest losses from spoilage and waste (Blades 1980). To try to correct for errors in reported changes in food stocks, the FAO uses the average food availability for three-year periods to calculate its undernourishment figures. This averaging does not address chronically biased measurement of production or spoilage.

The FAO is sensitive to these problems and their implications for mapping food insecurity. The organization acknowledges that errors in the estimate of available food can cause dramatic differences in reported patterns of undernourishment. For example, a 20% difference in the estimated food availability would be enough to alter the calculated undernourishment in a country from 7% to 64% of the population depending on the level of national food supply and inequality in distribution (Naiken 2003). As noted in the text accompanying map 2.1, reputable analysts consider errors in the measurement of availability of 25% to be quite possible.

Aside from estimating food availability, calculating the prevalence of undernourishment requires determining an individual's minimum daily caloric needs. The FAO uses data on the age and gender mix of a population to determine the minimum caloric requirement to support light activity for a "typical" person in a given country. The caloric cutoff is then set somewhat below the average requirement on the grounds that many people who eat less than the average requirement do so because they require less food based on their personal metabolism or activity level. The lowest cutoff established by the FAO in its *Sixth World Food Survey* (FAO 1996) was 1790 calories for South Asia. The highest cutoff was 1880 calories for East and Southeast Asia. While the cutoff is based on nutritional science, there is considerable debate about whether the estimated caloric requirements accurately reflect the needs of particular populations based on their activity levels, genetic makeup, and other factors (Smith 1998; Svedberg 1999). (See box 2.1 for examples of diets in this calorie range.)

Uncertainty about the appropriate cutoff point is troubling because the estimated prevalence of undernourishment is extremely sensitive to the cutoff point used in the calculation. For Sub-Saharan Africa, the FAO estimated the average per capita caloric requirement to

be 2100 calories. Based on this average requirement, the cutoff for Sub-Saharan Africa was set at 1800 calories. In 1991–92 the FAO calculated that 43% of the African population was undernourished. If the cutoff had been set 10% higher, at 1980 calories, 51% of the population would have been classified as undernourished. Had the cutoff been 10% lower, 34% of the population would have been labeled undernourished. Since existing nutritional studies could justify 10% increases or decreases in the cutoff, any estimate of undernourishment from one third to one half of Africa's population is equally valid under this method (Svedberg 1999).

Data on cutoff points and the national food supply are debated, but information on the distribution of food does not even exist in many countries. The FAO relies on a limited number of household food consumption and expenditure surveys to determine the distribution of available calories among households. It calculates the distribution of food at one point in time and assumes that distribution within countries has not changed for as long as 30 years. Some authors have noted that the method used to calculate undernourishment exaggerates the range of actual consumption levels, suggesting that some people eat unrealistically high (or low) amounts. Others contend that the method does not fully capture the degree to which poverty, as opposed to food supply, results in hunger.[8] In either case, the assumption that distribution is fixed over time means that change in food availability per person drives measured trends in undernourishment (Brown et al. 1995).

Even if there were flawless data, the FAO method of measuring undernourishment would still obscure important aspects of the hunger problem. This is because hunger may not be uniform within a household; its presence can come and go within a year; and it may exist even if sufficient calories are consumed.

### Individual Hunger versus Household Undernourishment

The prevalence of undernourishment as officially measured estimates the share of households without access to enough food to meet the caloric needs of their members. It does not measure the number of undernourished people directly. To come up with such a number, the FAO converts the percentage of households with inadequate food into estimates of the absolute number of people living in such households. But hunger is not necessarily shared equally in households. In many societies, for example, women are expected to "absorb" food shocks, and the particular nutritional needs of children are often neglected. Around the world, women are more likely to go hungry than men, while children are more likely to be undernourished than adults. Because nutrition is not uniform within households, the undernourishment calculation may be blind to many who are hungry in households that appear to have "enough."

### Seasonal versus Annual Data

The poor are often rural people who rely on farming for their food needs. Such people often have plenty to eat in the months immediately after harvests, but face hunger in the weeks before the next harvest. This seasonal hunger is not due to a failure to plan ahead. It is often caused by losses of stored food due to pests and spoilage or by unfavorable swings in market prices (low prices when farmers sell staple foods at harvest but high prices when farmers buy food before the next harvest) (Ferro-Luzzi et al. 2001).

Averaging national food supply over three-year periods means that hunger that arises only every few years or seasonally each year is not captured in the undernourishment indicator. For example, a hypothetical household in Africa whose members consumed 1900 calories each day for a year would not be recorded as undernourished because their consumption exceeds the 1800-calorie cutoff (though it is under the FAO's 2100-calorie average requirement). If they continued to consume this much for two years, but then went with no food at all for 30 days, the household would still appear to have no "undernourishment." The "extra" consumption over two years would have amounted to 73,000 calories per person, which is more than 1800 calories a day for 30 days. If the household's daily consumption returned to over 1800 calories per person for the rest of the third year, the average caloric intake of the household would exceed the cutoff. The members have obviously suffered food insecurity, if not hunger-related mortality, but the measure of undernourishment would not have recorded a problem.

### Malnutrition versus Undernourishment

In addition to caloric intake, people need protein and micronutrients to be well nourished. Failure to meet a threshold for caloric intake will leave a person hungry, but lack of specific nutrients such as iron, iodine, or Vitamin A can also have severe consequences for health, physical and mental development, and capacity to work. By focusing on calories, the undernourishment measure misses many forms of malnutrition. A household could enjoy food security in terms of the quantity of food consumed but lack nutrition security because of unbalanced diets (Benson 2004).

## UNDERNOURISHMENT AS A MEASURE OF HUNGER

The FAO measure of undernourishment is a better gauge of hunger than national food availability, but with so many possible sources of error, undernourishment may be overstated in some places and understated elsewhere. Consider these data for India and Ethiopia. For 2002–4 the FAO reported that 20% of India's population suffered undernourishment while 46% of Ethiopians were classified as undernourished. However, when nutritional deprivation is gauged by body mass index, which uses weight and height to assess whether a person is underfed, then 41% of Indian women and only 26% of Ethiopian women would be classified as suffering hunger (WHO 2004c). Both measures reveal an intolerable degree of food deprivation, but they give opposing impressions about where hunger is more widespread. They also imply large differences in the estimated total number of hungry people.

The undernourishment data are often used to indicate where hunger exists and how it is changing through time. Images like maps 3.1 and 3.2 are often presented to suggest general trends in the prevalence of hunger, progress toward meeting the Millennium Development Goals, and areas of the world that require particular attention. Since errors in measurement may not be consistent across regions, these maps may be misleading. For example, if food availability is consistently underreported in Africa, undernourishment in that region is overstated. Meanwhile, if gender discrimination causes particularly severe differences in nutrition within households in South Asia, hunger of women will be understated there. Despite its weaknesses as a measure of the number of hungry people, the FAO measure of undernourishment contributes to our understanding of the geography of hunger. Four useful functions of the undernourishment measure are discussed below with reference to maps 3.1 through 3.4.

*1. Identifying general, global patterns of hunger.* Undernourishment data are widely and effectively used to draw attention to global hunger and progress in its alleviation. These data are frequently used to cite the proportion of people suffering hunger (map 3.1), changes in that proportion (map 3.2), the total number of undernourished people in the developing world (map 3.3), and changes in that number (map 3.4). While these figures are indicative rather than precise, they clarify some aspects about global patterns in hunger. The exact numbers associated with map 3.3 are subject to error, but the impression that a large portion of the world's hungry people are found in India, China, and Central and Eastern Africa is reasonable. Similarly, from map 3.4 it is safe to conclude that efforts to reduce the number of hungry people have had more success in China and East Asia generally than in South Asia or Central and Eastern Africa. More specific conclusions

about the hunger situation are difficult to draw from the undernourishment data.

*2. Contrasting rates of undernourishment and numbers of undernourished.* Measuring hunger through "undernourishment" provides estimates of both the share of the population that is undernourished and the number of people in this condition.[9] Reference to both the prevalence and the total number of undernourished can be especially revealing in part because the two forms of the data sometimes leave very different impressions of progress. Presenting both the frequency of undernourishment and the total population that is undernourished can provide a nuanced view of the problem, but presenting one or the other can be misleading.

Because populations are growing, it is possible for the share of the population that is undernourished to fall (map 3.2), while the absolute number of people suffering undernourishment rises (map 3.4). For example, between 1980 and 2004, the estimated number of undernourished people in Bangladesh rose by about 11 million, from 33 to 44 million. Since the population of Bangladesh rose from 86 to 135 million, the share of the population that was undernourished fell from 39% to 30%. Table 3.1 and figure 3.1 give additional examples of countries that have experienced increasing numbers

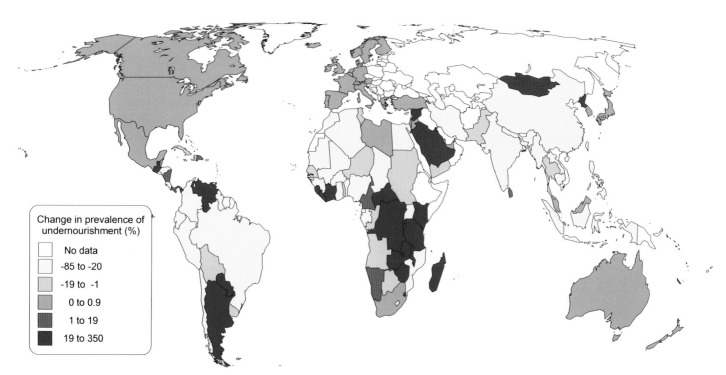

Change in prevalence of undernourishment (%)

| | |
|---|---|
| | No data |
| | -85 to -20 |
| | -19 to -1 |
| | 0 to 0.9 |
| | 1 to 19 |
| | 19 to 350 |

Map 3.2. Change in share of population that is undernourished, 1980–2004.

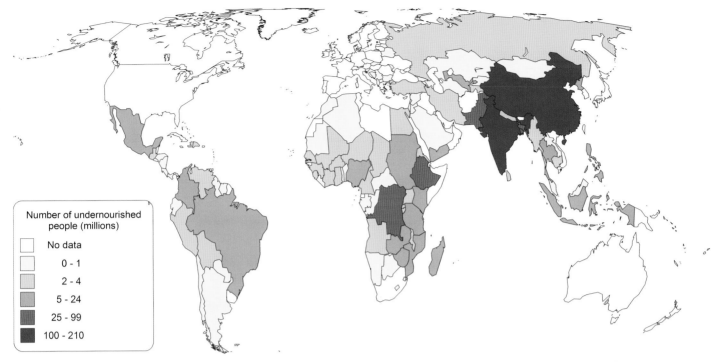

Map 3.3. Estimated number of undernourished people, 2002–4.

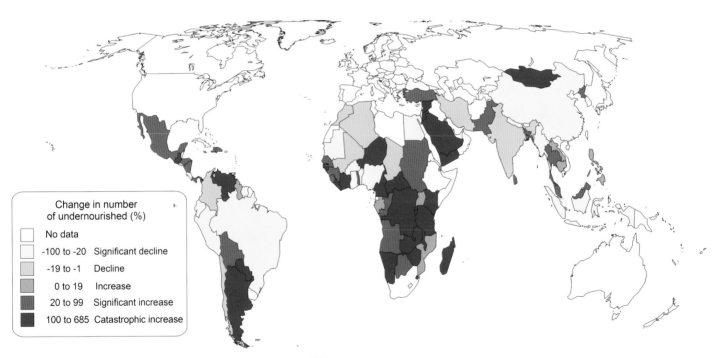

Map 3.4. Change in number of undernourished people, 1980–2004.

of undernourished people, despite declining rates of undernourishment. One lesson from the FAO measure of undernourishment is that consumers of data about hunger must think carefully about both rates and absolute numbers. Reductions in the prevalence of un-

dernourishment are certainly better than increases, but the goal of ending hunger requires a reduction in the absolute number of hungry people.

*3. Identifying hunger that is not caused by national food shortages.* Comparing the map of food availability

**Table 3.1. Undernourishment: Number (in millions) and Prevalence**

| | Number undernourished | | Change in number (%) | Percentage undernourished | | Change in percentage (%) |
|---|---|---|---|---|---|---|
| | 1979–81 | 2002–4 | | 1979–81 | 2002–4 | |
| Bangladesh | 33.3 | 44.0 | 32 | 39 | 30 | −23 |
| Pakistan | 23.6 | 37.5 | 59 | 29 | 24 | −17 |
| Cambodia | 4.0 | 4.6 | 15 | 60 | 33 | −45 |
| Laos | 1.0 | 1.1 | 10 | 33 | 19 | −42 |
| Philippines | 12.9 | 14.6 | 13 | 27 | 18 | −33 |
| Yemen | 3.2 | 7.6 | 138 | 39 | 38 | −3 |
| Uganda | 4.1 | 4.8 | 17 | 33 | 19 | −42 |
| Guinea | 1.5 | 2.0 | 33 | 32 | 24 | −25 |
| Togo | 0.8 | 1.2 | 50 | 30 | 24 | −20 |
| Sudan | 5.7 | 8.7 | 53 | 29 | 26 | −10 |
| Mozambique | 7.1 | 8.3 | 17 | 59 | 44 | −25 |
| Honduras | 1.1 | 1.6 | 45 | 31 | 23 | −26 |
| Bolivia | 1.4 | 2.0 | 43 | 26 | 23 | −12 |

Source: FAO 2008a, preliminary data for 2002–4.

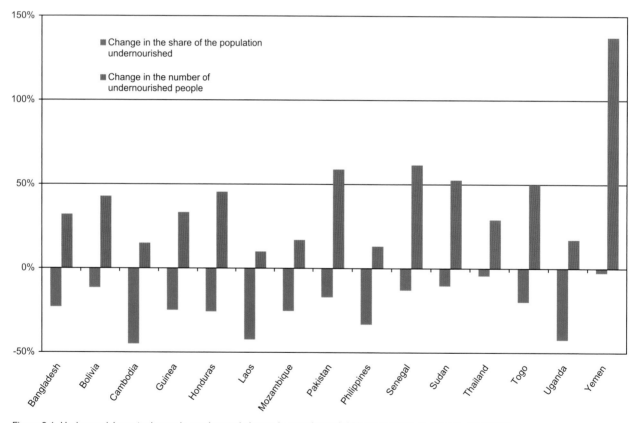

Figure 3.1. Undernourishment: change in number and change in prevalence (1979–81 to 2002–4). *Source:* FAO 2008c.

(map 2.1) with a map of undernourishment (map 3.1) suggests where food deprivation persists in the presence of abundance. We can see that the hungry in Brazil are victims of household poverty and its causes, not national food shortages. Indeed, food is more abundant in Brazil than in Uruguay, Chile, or Ecuador. But measured undernourishment is more prevalent in Brazil than in any of these other South American countries. The maps show that poverty alleviation and redistribution are more critical elements of a hunger strategy for Brazil than are efforts to raise food availability. In Africa, Nigeria stands out as a country where calories may be abundant, but distributional inequities result in undernourishment for millions of people. In much of the rest of Africa, both low food supply and poverty may be breeding hunger.

*4. Identifying worsening trends in hunger.* The FAO's measure of undernourishment is widely used to chart trends in hunger across the developing world. Due to the limitations in the data on food distribution, the measure is probably more accurate in capturing worsening trends in hunger than improvements. Because distribution is assumed to be constant in the FAO calculation, increases in food availability *always* imply reductions in measured undernourishment. Thus, countries with increasing availability of food in map 2.2 all show declining rates of undernourishment in map 3.2. In reality, increases in food availability probably alter food distribution. One would not expect the wealthiest households to consume more calories when more are available, since they already consume as much as they comfortably can. On the other hand, the very poorest households are also unlikely to consume very much of the additional food, because they are usually the last ones to see their capacity to buy food rise. Factors like low education, discrimination, and isolation which contribute to their poverty and hunger also tend to keep these households from enjoying income growth. It is

households that are poor, but not the poorest, that are most likely to experience increasing income and spend that income on food. If the members of these households had been consuming enough food to avoid undernourishment, their increased consumption might imply little reduction in hunger. Yet, the prevalence of undernourishment calculation suggests that hunger will fall with added food availability.

Increases in food availability may or may not bring reductions in hunger, but falling food availability probably does increase it. When average food supplies fall, the wealthy need not reduce food consumption and those who are already undernourished often cannot reduce food consumption further and survive. Thus, less food available per person probably means that reductions in food consumption are concentrated among the poor who might have previously consumed just enough to avoid undernourishment. Given reduced national food supplies, marginally nourished people become undernourished. With this in mind, examination of trends in undernourishment in map 3.2 suggests possible improvements in South Asia, where availabilities are growing, and almost certain worsening of hunger in Central and Eastern Africa, where availabilities are declining. Confirming the actual patterns in hunger over time and space requires more direct measurement of individuals.

# 4: Micronutrient Malnutrition

A diet for healthy living requires more than consuming a sufficient quantity of food. Healthy human development also requires the right quality of food. Beyond meeting a minimum caloric requirement, people need essential micronutrients like vitamin A, iron, and iodine for normal growth and development. People in wealthy countries usually get the micronutrients they need either through eating a diverse, balanced diet or as a result of fortification programs and dietary supplements. In contrast, it is estimated that globally two billion people, almost a third of the world's population, suffer physical or mental effects from micronutrient malnutrition. This represents more than two times the number of people thought to be suffering hunger from caloric undernourishment. Table 4.1 indicates the distribution of three particularly widespread deficiencies, which are mapped in maps 4.1–4.3.

Micronutrient deficiencies can have devastating effects on human health. They can impair mental development, reduce physical strength, and result in more frequent and more severe incidence of disease. Micronutrient deficiencies mean that each year hundreds of thousands of children go blind, hundreds of thousands

of women die in childbirth, and hundreds of thousands, perhaps millions, of people die from diseases that their immune systems could have protected them from had they been better nourished.

When people suffer caloric undernourishment, they almost inevitably suffer micronutrient malnutrition as well. For such populations, improving the quantity of food consumed helps to address deficiencies in vitamins and essential minerals. But data from the World Health Organization suggest that about a billion people consume sufficient quantities of food, but still suffer significant nutritional deficiencies. In these populations micronutrient malnutrition can be addressed through dietary supplements, food fortification, and dietary diversification. Recently, scientists have been developing crops to be richer in micronutrients. These biofortification efforts are discussed in boxes 4.1 and 4.2.

## VITAMIN A DEFICIENCY

The consequences of vitamin A deficiency (VAD) range from debilitating to deadly. VAD results in problems of vision and blindness, and impairs the body's defenses against infection. Through its effect on the immune system, vitamin A deficiency contributes to the high prevalence of infectious disease in developing countries and to high mortality rates from those diseases, including measles and respiratory or diarrheal infections (UNICEF 1998; WHO 2008d).

Over 125 million children under five years old are thought to suffer from vitamin A deficiency; 4.4 million are so severely deficient as to have eye damage. Up to a half a million of these children are expected to go blind annually, and two-thirds of those are likely to die of some infection within months of going blind (HarvestPlus 2003).

Animal products are rich sources of vitamin A, and the human body can synthesize the nutrient from beta-carotene, which is found abundantly in many fruits and vegetables. As map 4.1 shows, vitamin A deficiency remains most prevalent in Sub-Saharan Africa

**Table 4.1. Micronutrient deficiency rates by world region, 2002**

| Region | % of Population with deficiency in: | | |
|---|---|---|---|
| | Iron | Vitamin A | Iodine |
| Africa | 32 | 43 | 46 |
| Americas | 17 | 10 | 19 |
| Southeast Asia | 33 | 40 | 57 |
| Europe | n.a. | 57 | 10 |
| Eastern Mediterranean | 21 | 54 | 45 |
| Western Pacific | 14 | 24 | 38 |
| Total | 25 | 35 | 37 |

Notes: Vitamin A deficiency measured for children under the age of five.

Iodine deficiency measured through urinary analysis: UI < 100ug/L.

Iron deficiency measured from prevalence of anemia based on hemoglobin concentration.

Source: WHO 2008e.

**Box 4.1. Biofortification in Practice: Orange Sweet Potatoes**

UNICEF estimates that vitamin A deficiency (VAD) leads to 1.15 million child deaths a year. To address this health crisis, many countries support programs of micronutrient supplementation (distributing vitamins to children), industrial fortification (adding vitamin A to oil), and diet diversification. These efforts have improved nutrition for many people but not all. The very poor cannot afford to diversify their diet. Supplementation never reaches all children and can be prohibitively expensive if it is not done in the context of a larger health care program. Fortification fails to reach households that rely on home-produced foods and may send a mixed message about diet if it focuses on sugar or fats, foods that can have negative health implications in large quantities. To fill the gaps in providing micronutrients, biofortification aims to improve the health of poor people by increasing the nutritional quality of staple crops they grow and eat.

White cassava (sweet potato) is a staple for many people in Africa, but the varieties consumed there have few micronutrients. Meanwhile, yellow and orange varieties of cassava grown in South America are rich in beta-carotene, which our bodies converts into vitamin A. By crossbreeding the local varieties of white cassava with nutritionally superior varieties, scientists in Africa are developing biofortified "orange-fleshed sweet potatoes" that can be grown and consumed locally. If these new varieties are sufficiently rich in beta-carotene, if they produce good yields in farmers' fields, and if consumers accept them into their diets, then biofortified orange-fleshed sweet potatoes could be a powerful tool in reversing VAD in Africa. In Mozambique and Uganda mothers already have begun to feed their young children orange sweet potatoes, improving child nutrition (Van Jaarsveld et al. 2005).

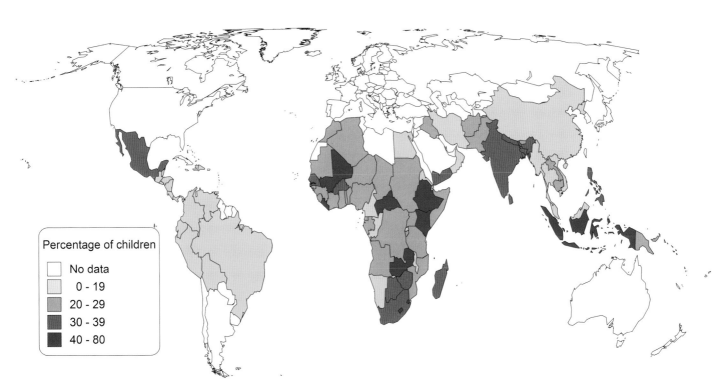

Percentage of children

☐ No data
▫ 0 - 19
▪ 20 - 29
▪ 30 - 39
■ 40 - 80

Map 4.1. Vitamin A deficiency in preschool-aged children. Most recent years.

and in South Asia. It is in the countries that have the most poverty and most caloric hunger that diets lack the diversity to deliver this needed nutrient. Alleviating poverty might address VAD. But the urgent problems associated with VAD need not wait for the end of poverty. Much can be done to address vitamin A deficiency among the poor right now.

Repeated experiments have found that twice-a-year massive doses of vitamin A reduce child mortality in poor countries by 25% to 30% (UNICEF 1998). Results

**Box 4.2. Biofortification in Practice: Golden Rice**

In contrast to cassava, no rice plant occurring in nature accumulates beta-carotene in the grain. Rice plants synthesize beta-carotene in the leaves and stem only. In the case of rice, biofortification was achieved through genetic modification rather than conventional breeding. Golden Rice is rice that has been genetically engineered to contain beta-carotene in the edible part of the grain. The concentration of beta-carotene in the grain results in a golden color, inspiring the name.

The first strains of Golden Rice were developed in 1999. Since then, varieties have been distributed to scientists in various parts of the world who are crossbreeding the genetically modified rice with local varieties to develop biofortified rice that can be grown and consumed in their countries. So far, no varieties have been released for widespread production. Slow progress from initial development to public distribution is partly due to concerns about the use of genetic engineering. These concerns relate to issues of ownership, human health, and environmental effects.[1]

*Ownership of technology.* When a new variety is developed through genetic engineering, there is the possibility that the distribution of the seed could be controlled by the developer to the detriment of the users. In the case of Golden Rice, the license to use the variety has been given to a humanitarian organization (Golden Rice Project) and is distributed free of charge. Licensing arrangements for Golden Rice are expected to allow farmers to save, consume, replant, or sell their resulting rice crop. Other crops developed through genetic modification could have more troubling ownership implications, such as requiring farmers to purchase seeds from a specific company each season.

*Human health and genetic modification.* Around the world, people worry that genetic modification could have unintended and negative health consequences for consumers. Some governments have prohibited distribution of genetically modified foods, while others have allowed it. There is no universal formula for deciding whether a genetically modified food is safe, but international organizations like the FAO, WHO, and OECD focus on comparison between the new food and foods that are currently considered safe. This type of analysis would probably favor acceptance of the ability of regulatory agencies to assess the safety of foods developed through these novel technologies. Those resistant to accepting genetically modified foods fear that the procedures used may be introducing hazards that are not yet recognized by science or consumers.

*The environment and genetic modification.* Genetically modified crops bring two main environmental concerns. The first is that the new varieties will become dominant, invasive weeds that are difficult to control. The potential for creating "superweeds" is particularly relevant when genetic modification intentionally develops herbicide-resistant varieties. This is not the case with Golden Rice. The second concern is that the genetically modified crop will transfer genes to wild species, affecting biodiversity in unplanned ways. Since Golden Rice has no genes that are not commonly found in plants and no traits that might cause it to dominate other species, it presents less of an environmental threat than crops that have been modified for other purposes or through the introduction of genes that are rarely (or never) found in plants.

The promoters of Golden Rice stress that the plant will not bind the poor into dependency on agricultural corporations, threaten the health of consumers, or alter our natural environment. Even if these points prove accurate, the potential for other genetically modified crops to have these effects remains. People and organizations who are cautious about genetically modified foods have expressed concerns that the introduction of Golden Rice, however beneficial, might open countries to many more genetically modified varieties and ultimately make inevitable the unintended consequences described above. Moreover, it remains to be seen whether Golden Rice can be grown by farmers or eaten by consumers to a large enough extent to contribute to better health. (Box 20.1 describes some of the problems in moving a new technology into use by farmers.) More fundamentally, some critiques view golden rice as reinforcing an emphasis on cereals production, when greater diversification of production into vegetables and other crops might provide better nutrition. In the long term, higher incomes and dietary diversification remain the most certain solution to micronutrient malnutrition. For the present, other interventions are needed if millions of people are to enjoy nutritious diets.

1. Detailed information on Golden Rice can be found through the Golden Rice Project Homepage (www.goldenrice.org). Much of the information presented here is from Golden Rice Project 2006. A less supportive assessment of this technology is found in Kirby 2003.

Figure 4.1. A fruit and vegetable seller and his assistants at a stand in the Shekawati area of Rajasthan, India. Photo by Kathleen O'Reilly.

of dosing with vitamin A supplements have been so dramatic that supplements have been added to child immunization programs in many countries. Through the 2002 National Immunization Day effort, vitamin A doses were administered to about 75% of the children in countries where VAD is known to be widespread. Such supplementation programs, however, may be phased out as diseases like polio are eradicated and immunizations are deemed unnecessary (Kennedy, Mannar and Iyengar 2003).

Fortification and biofortification are also being used to combat VAD. In Cuba and a number of other countries sugar and vegetable oils are enriched with vitamin A. Meanwhile crop breeders and scientists have developed varieties of rice, corn, and sweet potatoes rich in beta-carotene (boxes 4.1 and 4.2). These biofortification programs have been coupled with strong public education efforts to promote consumption of the new food varieties. A combination of supplementation, fortification, biofortification, and dietary diversification has helped combat vitamin A deficiency, but

the problem remains tenacious in the world's impoverished regions.

## IODINE DEFICIENCY

Iodine deficiency is the leading cause of preventable brain damage and mental disabilities in the world. An estimated 43 million people in the world suffer varying degrees of brain damage and physical impairment due to iodine deficiency, including 11 million who are afflicted with profound mental impairment. As many as 740 million people may have visible goiters, a swelling of the thyroid gland in the neck usually due to iodine deficiency (WHO/UNICEF/ICCIDD 2001, 17). Clinical evidence suggests that iodine deficiency that is not severe enough to cause goiter can lower intelligence quotient points by 10% (WHO 2004b).

The clinical indicator of sufficient iodine is 100 micrograms per liter (ug/L) of urinary iodine (UI). Iodine can enter the diet in sufficient quantities when seafood is commonly eaten, when soils are rich in iodine and impart it into cereals, and when fortification is

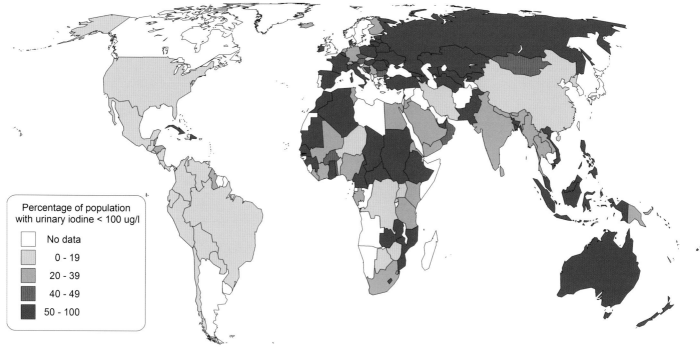

Map 4.2. Iodine deficiency.

practiced. Because environmental factors (like soil type and proximity to the sea) influence iodine deficiency, the geographic distribution of this nutritional problem shown in map 4.2 is very different from many others in this atlas. We see evidence of a problem in Central and Eastern Europe as well as in land-locked countries in Africa and Central Asia. At the same time, many countries with serious problems meeting other food needs have good consumption of this nutrient (e.g., Tanzania, India), and still others are at risk from excess consumption (Brazil, Chile).

Low-cost fortification programs have reduced the problem of iodine deficiency in many countries. Around the world campaigns to fortify salt with iodine have made tremendous progress. UNICEF reports that two-thirds of the world's population now has access to iodized salt. The World Health Organization estimated that the number of infants born with severe brain damage due to iodine deficiency dropped from 120,000 in 1990 to 55,000 in 2000 as a result of iodized salt (UNICEF 1998; WHO 2008d).

### IRON DEFICIENCY (ANEMIA)

Over two billion people are thought to be iron deficient (UNICEF 1998). The human body uses iron to produce hemoglobin, which is essential for moving oxygen from the lungs to tissues and taking carbon dioxide from body tissues to the lungs. Anemia, inadequate hemoglobin levels, is a well-known consequence of iron deficiency and results in reduced capacity for activity, lowering children's ability to learn and adults' ability to work. In more extreme cases anemia causes heart failure in young children and heart failure and death in pregnant women. Iron deficiency also results in developmental delays and behavioral disturbances in children, impaired immune responses, and fetal mortality. Beyond poor diet, anemia can be the result of parasites such as hookworms that deprive the body of nutrients after food is eaten (WHO 2008d).

Map 4.3 indicates the severity of iron deficiency among children in the developing world based on household surveys undertaken by the Demographic and Health Surveys.[10] The surveys show that anemia is a major public health problem in West and Eastern Africa and in South Asia. In these regions, a combination of low-quality diets and high prevalence of parasitic diseases demand multiple approaches to combating iron deficiency anemia.

Medicines to fight the parasites that rob people of nutrition are available for as little as three cents per per-

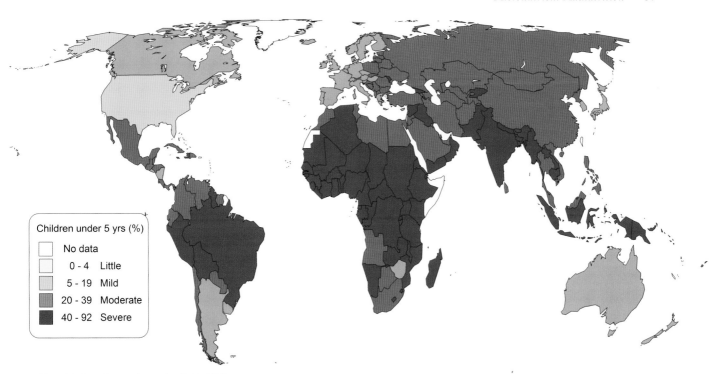

Map 4.3. Prevalence of anemia, 1993–2005.

son. Wheat flour is increasingly fortified with iron in Latin America and in the Middle East, and women in over 120 countries receive iron supplements supplied by UNICEF (UNICEF 1998). As with other micronutrients, balanced and diverse diets would go far in addressing iron deficiency anemia. However, the particular needs of pregnant women and the relationship between sanitation, parasitic infection, and anemia mean that even when people have sufficient income to diversify their diets, some public health interventions may be needed to combat iron deficiency anemia.

# 5: Malnutrition and Obesity

Malnutrition takes many forms. Previous maps focused on inadequate consumption of nutritious food among populations. The consumption of too much of certain foods can also result in malnutrition, but not hunger. When people do little physical activity and their diets include excessive amounts of saturated fats, starches, and carbohydrates, they become overweight or even obese. Nutritionists and health policy experts are increasingly concerned about this form of overnutrition because of its public health implications. Overweight and obese people are more prone to chronic diseases such as diabetes, hypertension, heart disease, osteoarthritis, and some cancers. Although the WHO and FAO focus their activities on ending hunger in the world, obesity is a growing concern (box 5.1).

The prevalence of obesity in a population is measured through body mass index (BMI) classifications developed by The National Center for Health Statistics (NCHS). A person's BMI score is his or her weight (in kilograms) divided by his or her height in meters squared. For example, a young woman who weighs 58 kilograms (128 lbs) and is 1.7 meters tall (5 feet 7 inches) has a BMI score of 20 ($58/(1.7)^2$). This number can be compared to the ranges developed by NCHS that classify subjects as underweight, normal weight, overweight, and obese (table 5.1). The ranges are indicative only of the frequency of these conditions in a population. Whether or not a specific individual's weight is a health concern should be determined in consultation with a physician.

Table 5.1 shows BMI cutoff levels that determine the NCHS classifications and presents the share of women in selected countries whose BMI scores fall into each of these ranges. NCHS nutritionists believe that people are at risk for diseases associated with undernutrition if their BMI is less than 18.5 and at risk of problems from overnutrition if their BMI is greater than 25. The use of a low BMI to indicate risks associated with undernutrition is described in box 5.2. With a BMI of 20, the young woman in our example would be considered to have a normal weight for her height. In contrast to this hypothetical woman, 52.2% of women in Turkey are classified as overweight with a BMI above 25. Almost 20% of the Turkish women surveyed were considered obese (BMI > 30). In Brazil, about 10% of women were obese, while 6% were underweight.

Like other indicators used in this atlas, the BMI measure has its strengths and weaknesses. First, the data are deficient in terms of the number of countries reporting. When data exist for a country, they are often for just one or two years. Poor geographical coverage and shallow time series make it difficult to chart global trends in obesity and related chronic diseases. Second, the cutoff points fixed by NCHS were developed for US citizens at high risk of disease above a BMI of 25 and below a BMI of 18.5. These standards and risks may not equally apply to other populations. When they are applicable, their relevance is limited to indicating tendencies in a population. The ranges and associated risks are not reliable diagnostics for individuals. We can think of examples that point to the limitations of BMI as a health indicator (e.g., a 6-foot-6-inch, 235-pound athlete with a BMI score of 27). There is consensus, however, that an adult BMI of greater than 30 kg/m$^2$ places individuals at considerable disease risk (WHO 2006a).

**Table 5.1. Women's body mass index percentiles for selected countries**

| | BMI < 18.5 Underweight | 18.5 < BMI < 25.0 Normal | BMI > 25 Overweight | BMI > 30 Obese |
|---|---|---|---|---|
| Bangladesh | 45.4 | 50.1 | 4.4 | 0.7 |
| Brazil | 6.2 | 58.9 | 34.8 | 9.7 |
| Burkina Faso | 13.2 | 81.0 | 5.7 | 0.9 |
| Egypt | 0.6 | 28.2 | 71.2 | 32.4 |
| Ethiopia | 26.0 | 71.6 | 2.3 | 0.2 |
| Guatemala | 2.0 | 54.1 | 43.8 | 12.1 |
| India | 26.6 | 69.7 | 3.7 | 1.0 |
| Kazakhstan | 9.8 | 67.5 | 22.8 | 8.4 |
| Turkey | 2.6 | 45.2 | 52.2 | 18.8 |
| Uzbekistan | 9.8 | 71.6 | 18.5 | 4.1 |

Source: WHO 2004a, Annex 11.

**Box 5.1. The Nutrition Transition and the Double Burden of Malnutrition**

Nutrition and health experts note that diets change significantly as a population becomes more urban and incomes rise. The "nutrition transition" refers to a shift from a predominately plant-based diet high in complex carbohydrates, fiber, and vegetables to an energy-dense (high calorie) diet in which a higher percentage of calories come from sugar, fat, and processed foods. For example, soft drink consumption has increased by 400% in Brazil since the mid-1970s and is believed to be a major contributor to obesity (Rohter 2005). Over the same period the population has shifted from 80% rural to 80% urban. The relatively lower cost of energy-dense foods versus vegetables, fruits, and lean meats partly explains the phenomenon of obesity among the urban poor (Drewnowski and Spencer 2004). Research shows that the income required to purchase fatty foods has decreased over the past 30 years (FAO 2006a).

The coexistence of underweight and overweight people in a population is not uncommon (Barquera et al. 2006). In Brazil, there are four million undernourished people but ten million who are obese. There is even evidence of malnutrition of both types within the same family. In one poor urban community in the Philippines, 8.2% of the households had an underweight child and an overweight mother. The percentage increased to nearly 20% in a more affluent urban community (Pedro and Benavides 2006). The "double burden of malnutrition" refers to the coexistence of undernutrition and overnutrition and their related infectious and chronic diseases. Government health programs must be versatile and well financed to prevent and control these twin nutritional problems.

**Box 5.2. Low BMI**

Just as a high BMI can indicate overnutrition, a low BMI suggests undernutrition. If a person has a BMI of less than 18.5, then he or she is considered underweight. Unfortunately, the data for low BMI are very limited and focus on adult women. Another problem with the data is that the BMI cutoff point of 18.5 may be too low for individuals whose weight fluctuates with food availability, work, illness, and income. A BMI of 20 would provide some

additional weight as a buffer for these lean periods (Svedberg 1999, 2087). Despite these data limitations, table 5.3 shows those countries with the highest levels of underweight women based on the low BMI measure. If we use this indicator alone, South Asia and Sub-Saharan Africa stand out as containing the highest percentages of malnourished women.

Data limitations make mapping the geography of obesity particularly challenging. The data that do exist are collected for children under the age of five and for women during their childbearing years. Map 5.1 shows obesity rates for children ages 0–5 based on the WHO's international child growth standards. The reference population for these standards is derived from surveys undertaken between 1997 and 2003 in six countries (Brazil, Ghana, India, Norway, Oman, and the United States). Obesity rates are lowest in Sub-Saharan Africa and South Asia and highest in East-Central Europe, the Middle East and North Africa. The countries with the highest rates of overweight and obese children are listed in table 5.2.

The United States has one of the highest adult obesity rates in the world (map 5.2). The Centers for Disease

Control and Prevention estimate that 30% of adults in the United States are obese (CDC 2008). This represents 60 million people or one-fifth of the world's 300 million obese adults. The prevalence of obesity is on the rise in the United States. In 1995 obesity rates were less than 20% in all 50 states. By 2005 there were just four states (Hawaii, Kansas, Connecticut, and Vermont) with prevalence rates less than 20%. Seventeen states recorded obesity levels greater than 25%, with three of them greater than 30% (Louisiana, Mississippi, and West Virginia). The obesity epidemic in the United States is associated with unbalanced diets and limited physical activity. The stereotype "couch potato" who sits in front of a video screen eating packaged snacks contains an element of truth about the reason for obesity in the United States. Obesity is also strongly associated with

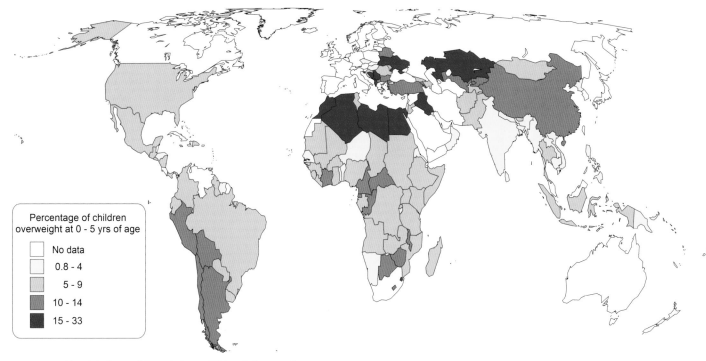

Map 5.1. Overweight and obese children 0–5 years of age.

**Table 5.2. Countries with the highest rates of overweight and obese children (BMI = +2SD)**

| Country | BMI |
|---|---|
| Albania | 33.0 |
| Ukraine | 27.6 |
| Bosnia and Herzegovina | 26.0 |
| Comoros | 25.3 |
| Serbia | 19.5 |
| Kazakhstan | 18.9 |
| Swaziland | 18.2 |
| Iraq | 17.1 |
| Algeria | 16.8 |
| Libya | 16.2 |
| Morocco | 15.5 |
| Egypt | 15.1 |

Source: WHO 2008c, most recent years.

**Table 5.3. Percentage of adult women underweight (BMI < 18.5)**

| Country | Rate (%) |
|---|---|
| Bangladesh | 45 |
| India | 41 |
| Eritrea | 41 |
| Nepal | 27 |
| Ethiopia | 26 |
| Yemen | 25 |
| Kampuchea | 21 |
| Chad | 21 |
| Niger | 21 |
| Madagascar | 21 |

Source: United Nations System, Standing Committee on Nutrition, 2004.

low income and low educational levels (Drewnowski and Spencer 2004). In many poor neighborhoods in US cities minimarts readily provide low-fiber, high-fat processed foods, but fresh produce and whole grain products are scarce and expensive.

The profile of obesity in the developing world is complex. In some countries, being overweight is a sign of wealth and high status. This is the case in urban Africa among the upper classes, and among "big men" in many rural African communities. In countries like Brazil and Mexico, it used to be prestigious to be over-

weight. While the elites of these countries are today adopting a slimmer look, obesity afflicts all socioeconomic groups (Rohter 2005).

In general, obesity tends to be higher in urban areas and is correlated with recent increases in a country's income (Martorell 2006). Urban populations in Latin America, Africa, and Asia are growing rapidly. As people move to cities, they encounter food that is high in sugar and fat and that is cheaper than more nutritious foods (Drewnowski and Spencer 2004; FAO 2006a) (figure 5.1). Poor people's access to nutritious

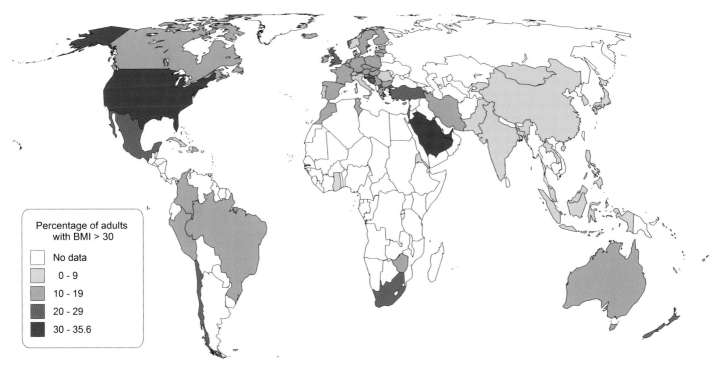

Map 5.2. Obese adults in the world.

Figure 5.1. A family eating fried chicken, french fries, and soft drinks at a fast-food restaurant in Tamarindo, Costa Rica. The consumption of high-fat foods and soft drinks combined with little physical exercise leads to weight gain and the onset of chronic diseases like diabetes.

food often becomes restricted if a limited number of stores sell a variety of fresh produce within walking distance of their homes (Algert, Agrawal, and Lewis 2006). Although people may know that eating more healthy foods is better for them, they cannot always afford to buy foods like fruits and vegetables (Drewnowski and Spencer 2004).

The relatively high prevalence of obesity in Latin America means that overnutrition commonly coexists with undernutrition (box 5.2 describes this phenomenon in Brazil). The combination of limited incomes, reduced physical activity, and an unbalanced diet increases the likelihood that poor children and adults alike will become obese and therefore be at risk for type-2 diabetes and hypertension, breast, colon, and prostate cancer, and heart disease (WHO 2006a). To counter the worldwide upsurge in obesity-related diseases, public policies need to promote the availability of low-fat, high-fiber foods among the poor, increase educational outreach and the opportunities for physical exercise, and support programs that address the links between income inequality, poor diets, and poor health.

# 6: Growth Failure

A simple but powerful indicator of hunger is the percentage of children in a population that suffers growth failure. Growth failure (or "stunting") is determined by measuring the height-for-age of children below the age of five. If a child's height is significantly below the average for his or her age, then growth failure is likely to have occurred. Growth impairment is linked to conditions of sustained poverty, especially chronic undernutrition and repeated illnesses that deplete a body's nutrients. Additional indicators of undernutrition include wasting and underweight. Wasting is defined as low weight-for-height. Wasting reflects current extreme deprivation, while growth failure indicates persistent chronic deprivation through early childhood. Underweight designates low weight-for-age.

The World Health Organization monitors children's heights and weights using surveys carried out by national governments. The WHO Global Database on Child Growth and Malnutrition compiles information from over 1700 national surveys representing 80% of the world's children under the age of five. How are the surveys conducted? Consider one example. The 2003 survey for Burkina Faso was based on a sample of 9381 children under the age of five. These children were weighed and measured to determine how close they were to an international reference group of children established by the WHO (box 6.1). Those whose height is well below the average for their age are considered to have short stature. More extreme divergence below the norm suggests severe growth impairment. Because of genetic variation, 2% to 3% of children would be classified as having short stature in a healthy, well-fed population. Many more children will have short stature if they are underfed. If more than 3% of the children have short stature, some growth failure is likely to have occurred. In 2003, 43.1% of children in Burkina Faso were classified as having short stature for age. Over 23% of these children were severely growth impaired. These high rates of short stature imply that many children experience growth failure due to poor nutrition.

Map 6.1 shows that short-statured children are found in all countries for which we have data. In the United States, 3.3% of children are of short stature, possibly due to genetic potential or possibly to health and nutrition. Unacceptably high rates of growth failure are found in all of the continents. The worst cases are in Sub-Saharan Africa, Central America and the Andes, and in South-Central and Southeast Asia. The countries reporting the highest growth failure rates are listed in table 6.1.

It would be useful to know whether rates of growth failure have been increasing or decreasing. The WHO provides data at the regional level for the period 1980–2005. Map 6.2 shows the changing proportion of children who have short stature within the UN-delimited regions over a twenty-five year period. All regions of the developing world show strong progress in reducing rates of growth failure, except Sub-Saharan Africa, where the trend is only slightly downward. Eastern Asia, for example, reduced the share of children with growth failure by 81% between 1980 and 2005. Map 6.3 tells a different side of the story. Using the same data as map 6.2 but focusing on changes in the *number* of short-stature children, the map shows that their number has significantly increased in most of Sub-Saharan Africa. In

**Table 6.1. Countries with the highest levels of growth failure (most recent years)**

| Country | Rate (%) |
|---|---|
| Burundi | 63.1 |
| Angola | 61.7 |
| Afghanistan | 59.3 |
| Niger | 54.8 |
| Guatemala | 54.3 |
| Madagascar | 52.8 |
| Malawi | 52.6 |
| Zambia | 52.5 |
| Rwanda | 51.7 |
| Yemen | 51.7 |
| Ethiopia | 50.7 |
| Papua New Guinea | 50.2 |

Source: WHO 2008c, most recent years.

**Box 6.1 Growth Failure and Growth Charts**

Figure 6.1 is a growth chart that is currently used in tracking the height-for-age of boys between birth and five years.

The X-axis is age from birth to five years and the Y-axis is height in centimeters. Once a child's height and age are determined, the height/age point is plotted on the chart. The green line in the middle of the chart corresponds to the height-for-age median value of the WHO reference population. This growth standard emerged from the WHO's Multicentre Growth Reference Study, which tracked the growth of more than 8000 breast-fed children in six countries (Brazil, Ghana, India, Norway, Oman, and the United States). The middle marks the 50th percentile of the reference group. A child whose height-for-age falls on that line is taller than 50% of the children that age and shorter than 50%. Growth failure is defined by the WHO as being two standard deviations or more below this median value. Many health professionals use the 3rd percentile on these charts as a cutoff point to indicate growth failure. They use it first because the line is drawn on the charts, and second because it is close to the −2 standard deviation indicator of moderate growth failure. If a 24-month old boy is found to be 81 centimeters tall, he would be identified as suffering growth failure. The WHO child growth standards replace the WHO/National Center for Health Statistics reference population which was based on a sample of children in the United States (WHO 2006b). The charts are prescriptive in that they identify the ideal growth curve of breast-fed babies whose mothers followed feeding recommendations and did not smoke (de Onis et al. 2007).

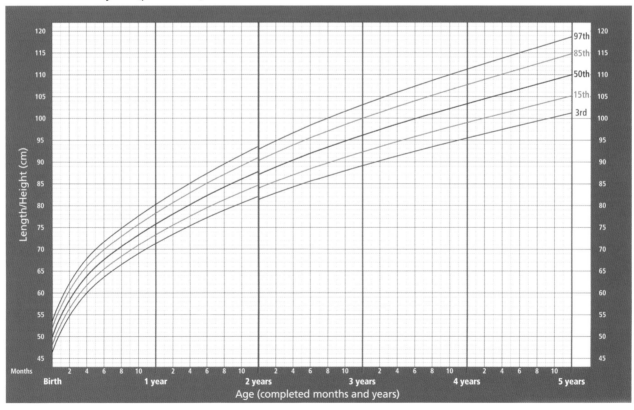

Figure 6.1. WHO growth chart for boys ages birth to five years. This chart is based on the new WHO standards introduced in 2006. It is derived from child growth studies in six countries and is used worldwide to monitor height-for-age of boys under five years.

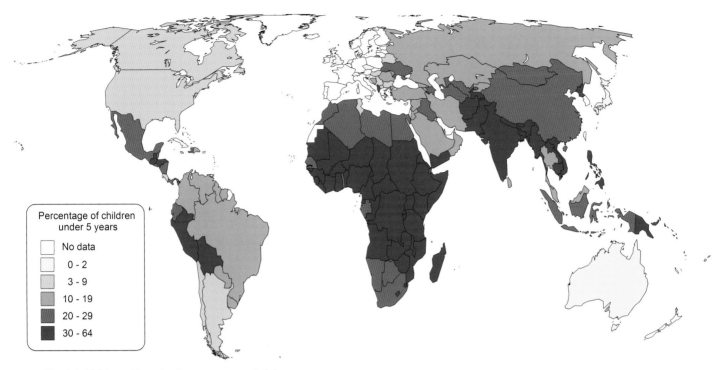

Map 6.1. Children with moderate or severe growth failure.

Map 6.2. Change in child growth failure rate, 1980–2005.

Map 6.3. Change in the number of children with growth failure, 1980–2005.

**Table 6.2. Children under five years experiencing growth failure, 1980 & 2005**

|  | Millions of children | | Percentage of children | |
|---|---|---|---|---|
|  | 1980 | 2005 | 1980 | 2005 |
| Eastern Africa | 12.0 | 21.6 | 44.4 | 44.4 |
| Middle Africa | 4.5 | 7.4 | 46.6 | 35.8 |
| Northern Africa | 6.2 | 4.2 | 34.0 | 19.1 |
| Southern Africa | 1.3 | 1.4 | 26.2 | 24.3 |
| Western Africa | 8.9 | 13.9 | 36.5 | 32.0 |
| Eastern Asia | 54.7 | 9.5 | 51.3 | 10.0 |
| South-Central Asia | 88.9 | 63.5 | 61.7 | 34.5 |
| Southeast Asia | 27.6 | 15.3 | 52.1 | 27.7 |
| Western Asia | 5.6 | 4.1 | 32.5 | 16.1 |
| Caribbean | 0.7 | 0.2 | 20.0 | 5.7 |
| Central America | 4.8 | 2.9 | 32.2 | 18.0 |
| South America | 7.2 | 3.4 | 21.3 | 9.6 |

Source: United Nations Standing Committee on Nutrition 2004

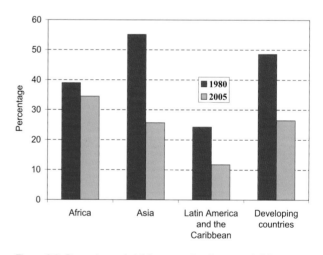

Figure 6.2. Percentage of children experiencing growth failure, 1980 and 2005. *Source:* UN Standing Committee on Nutrition 2004.

the middle Africa region, for example, the proportion of short-stature children in the population declined by 23% between 1980 and 2005 (from 46% to 35%), but the actual number of children increased by 64%.

Table 6.2 and figures 6.2–6.4 illustrate these trends in other ways. They show that despite progress in re-

ducing the incidence of growth failure in Eastern and Southern Asia, the largest number of short-stature children continue to reside there. Together they account for 60% of short-stature children in the world.

There are many advantages to using anthropometric indicators like growth failure to reveal the presence of hunger in the world. First, it is a relatively simple and specific indicator of individual malnutrition. Just a few variables are measured (height, weight, age, sex) and

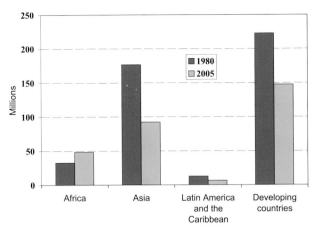

Figure 6.3. Number of children experiencing growth failure, 1980 and 2005. *Source:* UN Standing Committee on Nutrition 2004.

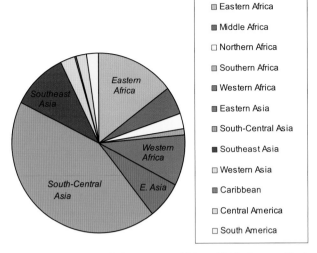

- ☐ Eastern Africa
- ■ Middle Africa
- ☐ Northern Africa
- ☐ Southern Africa
- ■ Western Africa
- ■ Eastern Asia
- ■ South-Central Asia
- ■ Southeast Asia
- ☐ Western Asia
- ■ Caribbean
- ☐ Central America
- ☐ South America

Figure 6.4. Number of short-stature children, 2005. *Source:* Black et al. 2008.

the potential for error is small. Health workers do not require extensive training, and their equipment is inexpensive and portable and can be easily obtained. Second, anthropometric measures give one insights into hunger at the scale of the individual. Food availability and undernourishment (maps 2.1 and 3.1) are national and household-scale indicators that do not necessarily portray the distribution of food among people. Growth failure, on the other hand, offers a flesh-and-bones

glimpse into levels of food consumption and nutrition within households (Shetty 2003, 143–44; Svedberg 1999, 2001).

Anthropometric measures do have their limitations. They tell us a lot about the status of children below the age of five but very little about adults, for whom little information is collected. Existing growth failure data

Figure 6.5. A group of children in Kericho, Kenya. One-third of the children under five experience growth failure in Kenya.

do not point to intrahousehold discrimination in nutrition among generations and between adult men and women. We need more inclusive surveys that consider all age and sex groups to gain insights into the distribution of hunger within households. Second, growth failure can occur as a result of illness rather than from insufficient food consumption. Intestinal parasites, infections that bring on fever, and diarrhea can sap a person's energy and nutrient levels, resulting in what is known as "secondary malnutrition." In isolation, growth failure rates cannot tell us whether malnutrition is due to health care and sanitation or inadequate access to food.

In summary, anthropometric measures like growth failure offer one of the clearest and least ambiguous views into the world of hunger. The maps show that South Asia and Sub-Saharan Africa contain the highest levels of hunger as revealed by the incidence of short stature among children. South Asia's prominence in the growth failure map (map 6.1) contrasts with its relatively high levels of food availability and moderate rates of undernourishment (see maps 2.3 and 3.1). Our effort to uncover hunger vulnerability as well as actual hunger continues in the next map, where we look at the relationship between poverty rates and hunger.

# 7: Household Poverty

It is poverty that prohibits most hungry people from getting the food they need. Unassisted households that lack the resources to buy or grow enough food to feed their members experience hunger. The poverty rate of a country can therefore indicate its vulnerability to food crises.

The poverty rate is the share of the population with incomes below an established poverty line. Typically, this poverty line is set at the income needed to meet a minimum food consumption standard, given other unavoidable expenses. For example, the first official poverty line for the United States was established in 1963 by estimating the cost of an economical food budget and multiplying that value by three. From country to country, ideas vary on what constitutes a minimum food consumption standard and an unavoidable expense. As a result, national poverty lines vary from country to country and tend to be lower in countries that have lower average expenditures (table 7.1).

**Table 7.1. National poverty lines**

| Country | National poverty line (dollars per day) (2005 PPP dollars) | Per capita expenditures (dollars per day) (2005 PPP dollars) |
|---|---|---|
| Albania | 2.83 | 9.33 |
| Argentina | 6.10 | 21.36 |
| Bangladesh | 1.03 | 2.13 |
| Benin | 0.80 | 2.43 |
| Bolivia | 4.73 | 7.20 |
| Brazil | 6.00 | 15.50 |
| Cameroon | 1.33 | 3.76 |
| China | 0.90 | 4.03 |
| Côte d'Ivoire | 1.37 | 3.90 |
| Ecuador | 4.07 | 9.66 |
| Ghana | 1.83 | 1.90 |
| India | 0.90 | 2.80 |
| Mexico | 6.40 | 21.03 |
| Mozambique | 1.00 | 1.50 |
| Tunisia | 1.37 | 8.03 |
| Uganda | 1.27 | 1.33 |
| Uruguay | 9.17 | 19.76 |
| Zambia | 1.33 | 2.00 |

Source: Ravallion, Chen, and Sangraula 2008.

To allow international comparisons and the measurement of global poverty over time, the UN has adopted the criteria for extreme poverty as consumption expenditures of $1.25 a day per person in 2005 "purchasing power parity" (PPP) dollars. This international poverty threshold is close to the national poverty lines of the 15 poorest countries in the world.[11] The $1.25/day limit is lower than most national poverty lines, as seen in table 7.1. The $1.25/day standard therefore yields a very conservative estimate of the total number of people in poverty, leaving out many people who would be considered extremely poor in their own countries. Recognizing this, the UN and other agencies also track data on households living on less than $2.00 a day. The $2.00/day threshold is the median poverty line for all developing countries (Ravallion, Chen, and Sangraula 2008, 12). In 2005 there were about 1.4 billion people living on less than $1.25 a day and over 2.5 billion living on less than $2.00 a day.

People living on less than $1.25 per day are defined as being in extreme poverty, but most people would agree that those living on $2.00 per day are also very poor. National poverty lines under $2.00 per day suggest that people living on that income can consume enough food to live. However, such people are highly vulnerable to hunger, since relatively small problems (like illness or theft) could force them to reduce their food consumption to an inadequate level. The $2.00/day poverty line captures households that are too poor to get enough food and also those that are at great risk of becoming extremely impoverished. It seems reasonable to assert that people living on $2.00 a day are at risk of falling into poverty even if they do not meet the strict poverty line set in their country. If the man buying a meal in figure 7.1 is a member of a household that earns less than $2.00 a day per person, he is probably at risk of hunger due to poverty, even if his income is above the national poverty line in Cote d'Ivoire ($1.37 per day).

Once a poverty line is established, data from house-

**Box 7.1. What Does $2.00/Day Poverty Mean?**

The statistic that over 2.5 billion people live on less than a $2.00 a day sounds too preposterous to be true. In reality, the $2.00/day poverty line does indicate people living in deprivation, but the "dollars" in the accounting are not US dollars. Rather, they are 2005 purchasing power parity (PPP) dollars. Conversion from US dollars to PPP dollars (also called "international dollars") is a simple process that reveals what it means to live in $2.00/day poverty.

Most countries keep track of people's incomes, measuring them in local currency units (pesos in Mexico, euros in France). To make comparisons of poverty across countries, incomes and expenditures need to be converted into a common currency. One could make this conversion using market exchange rates, but to do so would be misleading if the costs of living differ widely among countries. The use of 2005 PPP dollars addresses the problem of differences in cost of living across countries and also corrects for inflation over time.

To understand the PPP figures, imagine that in 2005 the cost of securing a 2100-calorie diet in Kenya was 32 Kenya shillings (Ksh). Suppose that the same diet could be purchased for US$1 in the United States. Since Ksh32 buys the same consumption as US$1, the purchasing power parity exchange rate is Ksh32 to US$1. Kenyans living on less than Ksh32 per day were below the $1(PPP)/day poverty line. Those with less than Ksh64 per day were in $2(PPP)/day poverty.

Purchasing power parity exchange rates are likely to be very different from actual market exchange rates. Indeed, in 2005 the PPP exchange rate between Kenya shillings and US dollars was about Ksh32 to $1(PPP), but the market exchange rate was Ksh75 to US$1. The difference between these rates means that Americans in Kenya found food prices very low. One US dollar bought Ksh75 at the banks, but only Ksh32 was needed to buy the food that one dollar bought in the US. If market exchange rates were used to calculate the poverty line, Kenyans earning less than Ksh150 would be considered to be in $2.00/day poverty. Reflecting the lower cost of living in Kenya, PPP exchange rates are used, and the $2(PPP) poverty threshold is Ksh64. Those 64 Kenya shillings bought about the same amount in Kenya in 2005 as two US dollars bought in the United States in that year.

Because of inflation a US dollar buys less each year. In 2007, for example, consumers in the United States needed $1.06 to buy what had cost $1.00 in 2005. It would be misleading to suggest that a person living on $1.06 in current US dollars in 2007 was better off than one living on US$1 in 2005. To avoid comparison of incomes measured in 2005 dollars with incomes measured in 2007 dollars, poverty data are tracked using indexes to adjust for inflation. A person living on $2.00 (2005 PPP) can consume as much as a US consumer could buy for US$2 in 2005.

Figure 7.1. A food vendor serving a meal in an open-air restaurant in Abidjan, Côte d'Ivoire, July 2005. Photo by Moussa Koné.

hold surveys are used to assess how many people live in poverty. These surveys record either the expenditures of a household or the income of the household over a period of time and then divide that value by the number of people in the household to estimate how much money each person lives on per day. Once information on consumption per person is gathered, it is converted from local currencies (like pesos in Mexico) into US dollars using special conversion factors that correct for differences in prices between countries. These conversion factors are called purchasing power parity (PPP) exchange rates. If correctly calculated, $1.00 (PPP) can buy the same bundle of goods in any country as US$1 can buy in the United States. PPP conversions are explained further in box 7.1.

Cutting the share of the population living in extreme poverty to half of its 1990 level is central in the UN's Millennium Development Goals. As figures 7.2 and 7.3 show, progress is being made toward this goal. Globally both the number of people and the share of the population living on less than $1.25 per day have declined over the last 25 years. The drop has been most dramatic in East Asia, where the number of people living in extreme poverty fell from over a billion in 1981 to 873 million in 1990 and 316 million in 2005. The share of East Asia's population living on less than $1.25 a day dropped from 78% in 1981 to 17% in 2005.

Headway in combating extreme poverty has been geographically uneven. Figure 7.3 indicates that the number of people in extreme poverty has actually risen in Sub-Saharan Africa and in South Asia. Despite worldwide declines between 1981 and 2005, the number of extremely impoverished people in Africa rose from 214 million to 391 million. More than half of that

continent's population lived on less than $1.25 per day during this time.

Many people who have escaped extreme poverty remain vulnerable. Indeed, the decline in the number of people living on less than $1.25 a day was not matched by a fall in $2.00/day poverty. As figure 7.4 shows, the number of people living on less than $2.00 a day rose from 2.5 billion in 1981 to 2.7 billion in 1990 and then returned to just above 2.5 billion in 2005. Only in East Asia do we see a decline in the number of people living below $2.00 a day. The share of the population living on less than $2.00 a day has fallen in most regions, but it has shown little change in Africa and in Europe and Central Asia (figure 7.5). Overall the $2.00/day poverty rate has fallen, but this accomplishment should not blind one to the reality that more people lived below this poverty threshold in 2005 than in 1981.

Declining poverty rates in East Asia and growing poverty in Africa have changed the global distribution

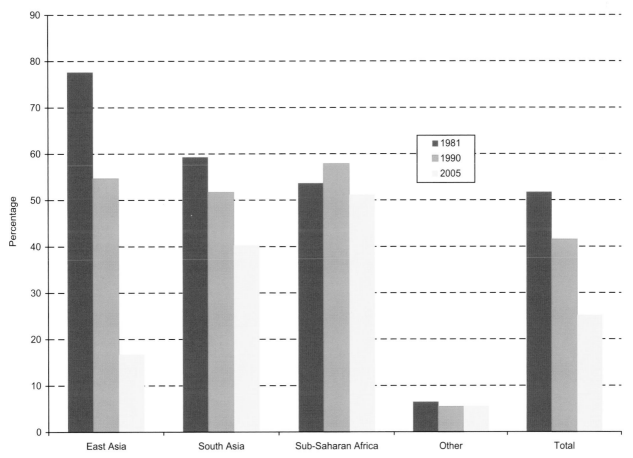

Figure 7.2. Percentage of population living on less than $1.25 per day, 1981, 1990, 2005. *Note:* "Other" includes Europe, Central Asia, Latin America and the Caribbean, the Middle East, and North Africa. *Source:* World Bank, PovcalNet, n.d.

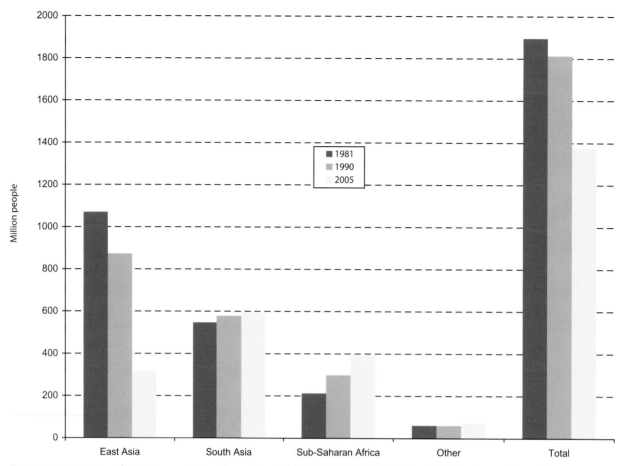

Figure 7.3. Distribution of $1.25/day poverty (in millions of people), 1981, 1990, 2005. *Note:* "Other" includes Europe, Central Asia, Latin America and the Caribbean, the Middle East, and North Africa. *Source:* World Bank, PovcalNet, n.d.

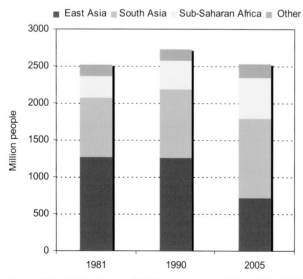

Figure 7.4. Distribution of $2.00/day poverty, 1981, 1990, 2005. *Note:* "Other" includes Europe, Central Asia, Latin America and the Caribbean, the Middle East, and North Africa. *Source:* World Bank, PovcalNet, n.d.

of the poor. As the pie charts reveal (figure 7.6), almost half of the world's poor were located in East Asia in 1981. By 2005, a much larger share of the world's poor resided in South Asia and Africa. Maps 7.1 and 7.2 reveal this concentration of high $2.00/poverty rates in Sub-Saharan African and South-Central Asia.

The relationship between poverty and hunger is reflected in figure 7.7. This illustration plots each country in terms of its rate of poverty and its rate of undernutrition as measured by child growth failure. The scatter diagram shows a fairly clear pattern in which countries with low poverty rates also have low rates of growth failure. Countries with high poverty rates tend to have high rates of growth failure. Exceptionally low rates of growth failure for a country with high poverty rates could reflect government policies and entitlement programs aimed at improving child nutrition and health care. Deviations from the general pattern of increas-

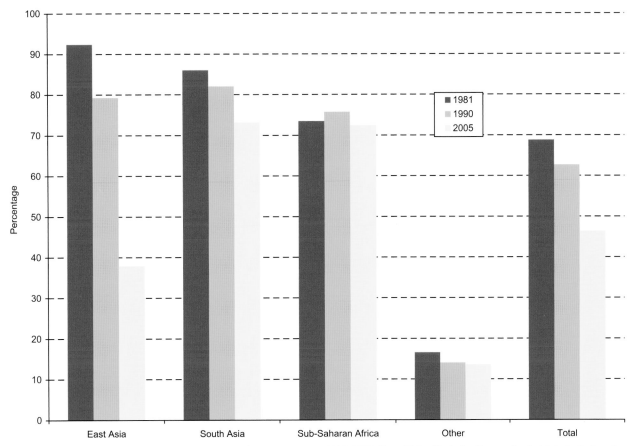

Figure 7.5. Percentage of population living on less than $2.00 a day, 1981, 1990, 2005. *Note:* "Other" includes Europe, Central Asia, Latin America and the Caribbean, the Middle East, and North Africa. *Source:* World Bank, PovcalNet, n.d.

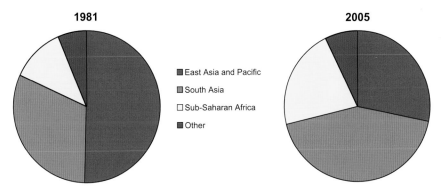

Figure 7.6. Distribution of population living on less than $2.00 a day, 1981 and 2005. *Note:* "Other" includes Europe and Central Asia, Latin America and the Caribbean, the Middle East and North Africa. *Source:* World Bank, PovcalNet, n.d.

ing undernutrition with increasing poverty could also reflect errors in measuring the poverty rate or the incidence of short-stature children.

Poverty corresponds closely to undernutrition, but there is also a strong correlation between poverty and national food availability (figure 7.8). Having an adequate supply of food in a country does not ensure

that households will have access to that food, but figure 7.8 reveals a tendency for countries with low food availability (under 2200 calories per capita) to also have very high poverty rates (over 45%). Figure 7.8 is divided into four quadrants. Quadrant A contains countries with low food availability and high poverty rates. In these countries, access and availability are both serious

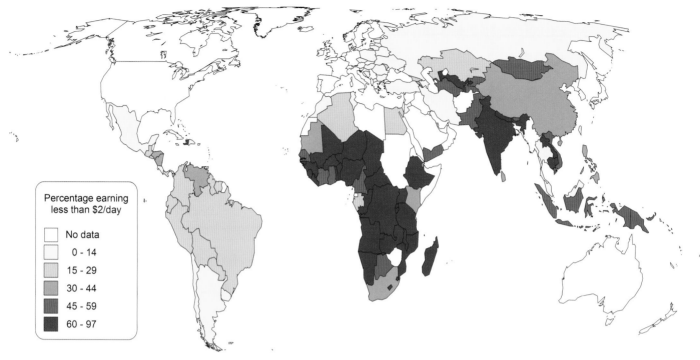

Map 7.1. Share of population living on less than $2.00 a day.

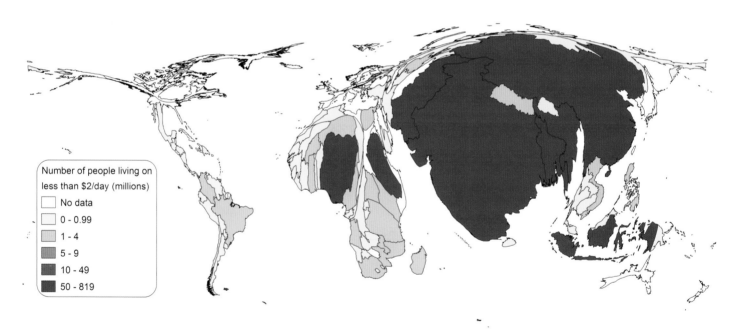

Map 7.2. Number of people living on less than $2.00 a day.

problems. Countries in quadrant B have high poverty rates and high national food availability. In these countries hunger is more a matter of unequal access to food than its supply. As one would expect, there are very few countries in quadrant C with low food availability and low poverty. Countries with low poverty generally im-

port or produce enough food to keep availability high. Finally, quadrant D contains countries with low poverty rates and high food availability.

Countries with low poverty and high food availability are unlikely to have as great a hunger problem as those that have high food availability combined with

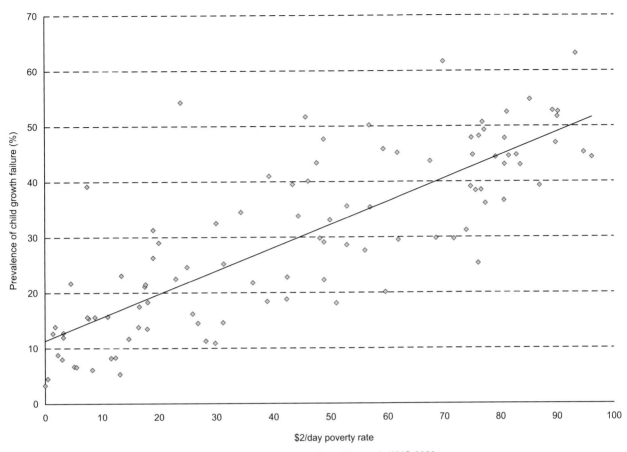

Figure 7.7. Poverty rate and child growth failure. *Sources:* World Bank, PovcalNet, n.d.; WHO 2008c.

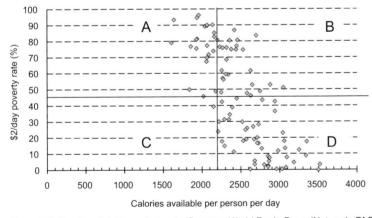

Figure 7.8. Food availability and poverty. *Sources:* World Bank, PovcalNet, n.d.; FAO 2008b.

high rates of poverty. Figure 7.9 presents the relationship between child growth failure and poverty rates in countries that have adequate food availability (those in quadrants B and D in figure 7.8). As the scatter plot shows, countries with higher poverty rates also have higher rates of growth failure, even when they have adequate food availability.

Poverty can be measured in many different ways. Those who must manage on less than $1.25 per day live in poverty so extreme that basic requirements for health and nutrition cannot be met. Over the last 20 years the total number of people facing these conditions has fallen globally, but it has risen in Africa. People who live on $2.00 a day may not be deprived of basic

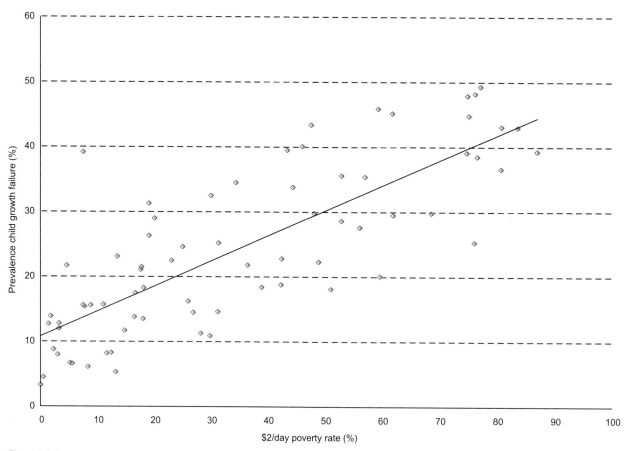

Figure 7.9. Poverty and growth failure in countries with adequate food availability (over 2200 calories per day per person). *Sources:* World Bank, PovcalNet, n.d.; WHO 2008c.

needs, but they are constantly at risk of falling into extreme poverty. They are at risk of hunger. Over the last 25 years, the share of the population experiencing this type of poverty has fallen in every major region, but the total number of people living on less than $2.00 a day has risen. The distribution of child undernutrition, measured by growth failure, tends to confirm the assertion that $2.00/day poverty corresponds with hunger. Regardless of total food availability, greater prevalence of $2.00/day poverty implies greater tendency for growth failure in children.

# 8: The Hunger Vulnerability Index

Widespread hunger can emerge when the national food supply is insufficient, when poverty keeps households from accessing available food, and when individuals cannot benefit from the food that their household controls due to distribution, disease, or other factors. The measures of hunger presented so far each address different levels of the hunger problem. Food supply is captured in food balance sheets; household access is gauged by the $2.00/day poverty rate; and individual nutritional outcomes are reflected in child growth failure. Another way to describe a country's hunger problem is to combine food availability, household access, and nutritional outcomes into a single index. This approach yields a summary measure of vulnerability to hunger.

The Hunger Vulnerability Index begins with food availability. In countries that have an average daily availability of over 2300 calories per capita, the existence of hunger in the population cannot be considered a result of inadequate total supply. The degree to which availability may be driving hunger should be reflected in the difference between measured food supply and this threshold. The first element of the Hunger Vulnerability Index is simply the percentage by which food availability falls short of 2300 calories per day. For example, with only 1960 calories available per person per day in Tanzania, that country has an availability score of 2300 minus 1960 divided by 2300, or 340/2300 = 0.15. Expressed as a percentage, the availability component of Tanzania's hunger index is 15. For countries with food availability over 2300 calories per day, like Brazil and India, the availability component of the index is zero.

Access to food at the household level is determined largely by the prevalence and severity of poverty. The higher the prevalence of poverty in a country, the greater the vulnerability of its population to hunger. People who live in higher income households (figure 8.1) are far less vulnerable than those who live in poverty. The exact income required to ensure household food security varies from place to place, but an international poverty line of $2.00 per day provides a criterion for assessing those at risk of hunger (map 7.1). The Hunger Vulnerability Index takes the share of the population living on incomes below this threshold as a measure of the degree to which household access to food places people at risk of hunger. Thus, in Tanzania, where 96% of the population has less than $2.00 per day, the access component of the hunger index is 96. With 18% of its population living on less than $2.00 per day, Brazil has a score of 18, while India has a score of 75.

The nutritional status of individuals is best measured

Figure 8.1. A Nepali cook making tea in an upper-income household in the Shekawati area of Rajasthan. Photo by Kathleen O'Reilly.

**Box 8.1. Hunger Vulnerability Diamonds**

The HVI diamonds in figure 8.2 illustrate how hunger vulnerability varies in its scale and form across countries. These diamonds plot the scores for availability, access, and nutritional outcome as well as the HVI. The points of the diamond are the highest score of any country for which there is data. In each case a higher score represents greater hunger vulnerability. The area of each country's diamond suggests the size of the hunger problem. In the cases plotted, Brazil's relatively small diamond indicates a smaller problem there than in India or Tanzania. Tanzania and India have extremely high HVI scores, indicating serious hunger problems, but the diamonds for these countries have very different shapes. In India, availability does not appear to be a factor in the hunger problem, while access and outcomes dominate. The "diamond" for India is in fact a triangle, since availability is met, but problems exist in access and in nutritional outcomes. In India and Brazil, the "diamonds" are almost equilateral triangles, suggesting that fighting poverty and preventing growth failure outcomes are both high priorities in these countries. In Tanzania, all three possible sources of failure come into play. Hence, the diamond for Tanzania has four sides. In Tanzania's case, the shape is skewed toward poverty, indicating that addressing access to food may be particularly critical there. The large role of availability implies that efforts to lower poverty while raising food production could be particularly valuable. The diagram shows that hunger varies not only in its scope and intensity, but also in its source.

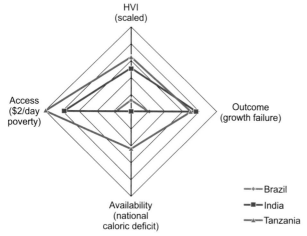

Figure 8.2. The shape and size of the hunger problem. *Sources:* WHO 2008c; FAO 2008b; World Bank, PovcalNet, n.d.

**Table 8.1. The Hunger Vulnerability Index: Selected countries**

| Country | National availability (food imbalance) A | Household access (poverty rate) B | Individual outcome (growth failure) C | Hunger Vulnerability Index D |
|---|---|---|---|---|
| Armenia | 2 | 42 | 19 | 21 |
| Bolivia | 4 | 29 | 32 | 22 |
| Brazil | 0 | 18 | 14 | 11 |
| China | 0 | 36 | 22 | 19 |
| Ethiopia | 19 | 77 | 51 | 49 |
| India | 0 | 75 | 48 | 41 |
| Indonesia | 0 | 53 | 29 | 27 |
| Kenya | 7 | 39 | 41 | 29 |
| Pakistan | 0 | 59 | 46 | 35 |
| Tajikistan | 20 | 50 | 33 | 34 |
| Tanzania | 15 | 96 | 44 | 52 |
| Uruguay | 0 | 2 | 14 | 5 |

Note: The Hunger Vulnerability Index is the sum of columns A, B, and C divided by 3.

Sources: See sources for maps 2.1, 6.1, and 7.1 in appendix 1.

anthropometrically. Because it is the most widely available and relevant measure, the share of children with clinical short stature is used in this index. One would expect low rates of child growth failure in countries where few people are vulnerable to hunger and higher rates as vulnerability becomes more common. In Tanzania, 44% of children under the age of five have short stature, implying a nutritional outcome score of 44. Child growth failure rates in Brazil and India are 14% and 48%, respectively.

Insufficient national food supply, prevalence of household poverty, and rate of child growth failure each reflect a level at which nutritional problems can emerge. Giving each level of the hunger problem equal weight, one can take the simple average of these three components to calculate a Hunger Vulnerability Index (table 8.1). This composite index must fall between zero and one hundred. To score zero, the country would have a food balance of over 2300 calories, no people below the $2.00/day poverty line, and no short-stature children. To have a score of one hundred, the country would have to have absolutely no food, all people earning less than a $2.00 per day, and all children experiencing growth failure. Since this upper limit is impossible, it is sometimes useful to scale the rates of food insufficiency, poverty, and growth failure by dividing them by the maximum recorded rate before taking the average. This scaling is used in figure 8.2.

As long as a country has a high score on any of these components, its population is vulnerable to hunger and the index will reveal this vulnerability. Examination of the components of the index can suggest whether the problem emerges at the level of national availability, household access, individual use, or some combination of them. Because the index combines the three aspects of the hunger problem, it can show how countries with different conditions at each level may be equally vulnerable to hunger. For example, table 8.1 shows that Pakistan and Tajikistan differ significantly in terms of food availability (a problem in Tajikistan but not in Pakistan) and child growth failure (a more common problem in Pakistan). Use of either of these measures would suggest very different hunger problems in these two countries, but the hunger index points to similar levels of hunger vulnerability. This is a reasonable result. High rates of growth failure in Pakistan reveal that a problem exists, despite measured availability. Meanwhile low availability and widespread poverty leave Tajikistan's population more vulnerable to hunger than its rate of growth failure may suggest.

The index of vulnerability can reveal differences among countries that are obscured by isolated consideration of availability, poverty, or growth failure. Bolivia, for example, has a similar rate of growth failure as Tajikistan, but there is much lower overall vulnerability in Bolivia because poverty rates are lower and food availability is greater. Meanwhile, Bolivia and Armenia have similar rates of hunger vulnerability. In Armenia, however, health outcomes in terms of growth failure are much better than in Bolivia. Armenia's vulnerability results from its high poverty rate, which leaves many people at risk of hunger, even if they are currently avoiding growth failure.

The appeal of using this index is that it can provide a summary impression of the hunger situation that is not always reflected in the individual measures. This composite measure is also useful for identifying a

problem. Understanding the problem can begin with examination of the components of the index, but must also include consideration of the many causal factors identified in figure 1.2. In contrast to the FAO measure of undernourishment, this index does not purport to measure the number of hungry people. It can, however, indicate how the hunger problem differs across countries and how it is changing over time. Table 8.2, for example, reveals the progress made in China at reducing hunger vulnerability in the 1990s. While food availability has been above the threshold throughout the period, reductions in poverty and child growth failure have generated a steady reduction in the hunger index.

As with any measure of the global food situation, the Hunger Vulnerability Index has its weaknesses. First, the simple averaging of the three components arbitrarily gives equal weight to each of them. A 1 percentage-point reduction in any element of the index creates the same decline in the HVI. This weighting is defensible on the grounds that success at each scale of the hunger problem is essential to its solution, but other arguments might emerge. Second, erroneous

**Table 8.2. Hunger Vulnerability Index: China, 1990–2005**

| Year | National availability (food imbalance) | Household access (poverty rate) | Individual outcome (growth failure) | Hunger Vulnerability Index |
|---|---|---|---|---|
| 1990 | 0 | 84 | 31* | 38 |
| 1992 | 0 | 78 | 31 | 36 |
| 1994 | 0 | 78 | 30* | 36 |
| 1995 | 0 | 72* | 29 | 34 |
| 1996 | 0 | 64 | 25* | 30 |
| 1998 | 0 | 61 | 16 | 26 |
| 2000 | 0 | 56* | 14 | 23 |
| 2001 | 0 | 51 | 14 | 22 |
| 2005 | 0 | 36 | 22 | 19 |

Note: Food availability rose in China from 2327 calories in 1980 to 2709 in 1990 and 2969 in 2000. This growth in availability produces a reduction in the FAO measure of undernourishment that is similar to the one recorded by the Hunger Vulnerability Index. While both measures record a reduction in hunger, the HVI attributes that decline to reduced poverty and ignores increases in availability above 2300 calories, whereas the FAO measure responds to the increased availability and ignores changes in poverty and nutritional outcomes.

*Value imputed from reported data.

Sources: See sources for maps 2.1, 6.1, and 7.1 in appendix 1.

**Box 8.2. Which Index to Measure Hunger?**

Because the origins of hunger are complex, it is useful to blend measures of different aspects of food insecurity into a single index. However, there are many variables to choose from and many ways they could be combined. In addition to the Hunger Vulnerability Index (HVI), there are at least two other indexes currently in use: the Global Hunger Index (GHI) and the Poverty and Hunger Index (PHI). Each of these indexes is based on slightly different data to measure hunger with slightly different results. Variables in the HVI highlight factors placing people at risk of hunger. The GHI emphasizes experience with hunger, and the PHI focuses more on extreme poverty.

*Hunger Vulnerability Index.* One can think of the hunger status of a country as a product of national food availability, access to food among households, and the nutritional outcomes of individuals. The Hunger Vulnerability Index captures national food availability by measuring deficiency in total food supplies in a country; it measures household access to food through poverty rates based on a $2.00/day poverty line; and it measures individual nutritional outcomes through the frequency of child growth failure. The HVI gives equal weight to national availability, household access, and individual outcomes. This weighting reflects the idea that a worsening in any one of these dimensions of hunger would place more people in a country at risk of malnutrition (map 8.1). The other indexes measure availability, access, and outcomes with different data and give different weight to each of these factors.

*The Global Hunger Index.* The Global Hunger Index (GHI) was introduced by the International Food Policy Research Institute in 2006 to draw attention to hunger problems and to provide a means of tracking progress in the fight against hunger (Weismann 2006). The GHI gives equal weight to the prevalence of undernourishment in a country, the share of children under five who have low weight for their age, and the rate of mortality among children under five. The prevalence of undernourishment, as calculated by the FAO, is a composite of national food availability and the distribution of that food supply across households. Thus one component of the GHI is a joint measure of access and availability. The other two components of the Global Hunger Index (weight-for-age and child mortality) can both be considered to reflect the nutritional outcomes of individuals. Thus, compared to the HVI, the GHI gives more emphasis to individual outcomes and less to access and availability.

In general the GHI and the HVI give similar impressions of the world hunger problem. There are, however, many countries for which the HVI indicates high hunger vulnerability, but the GHI suggests less concern. Of the 43 countries for which the Hunger Vulnerability Index indicates a very high or extremely high hunger problem, only 31 are ranked as alarming or extremely alarming by the GHI; the other 12 are placed in the GHI's middle category (serious hunger). Of the 16 countries that fall in the middle range of the HVI classifications (high hunger vulnerability), 13 are ranked in the middle GHI class (serious hunger) and 3 are ranked in its low or moderate hunger range. Table 8.3 looks at some of the countries for which the HVI and GHI score contradict. In most cases, the HVI is higher than the GHI because there are relatively low rates of child mortality or growth failure but a large share of the population that lives on less than $2.00 a day. These are countries where nutritional and health programs or other safety nets may be compensating for poverty. Arguably, the populations remain vulnerable to hunger if the health programs are vulnerable. If these programs are secure, populations are less vulnerable. As long as neither the GHI nor the HVI is seriously affected by poor data quality, the GHI identifies countries that experience poor nutritional outcomes, while the HVI reveals countries where the population is at risk of hunger.

*Poverty and Hunger Index.* As the name implies, the Poverty and Hunger Index (PHI) places more emphasis on poverty. It is published by the World Food Programme to monitor progress toward achieving the Millennium Development Goal of cutting hunger and poverty in half between 1990 and 2015 (World Food Programme 2006a). Because the Millennium Development Goals initially defined poverty in terms of the $1.00/day threshold and hunger in terms of undernourishment and low weight-for-age, these variables are used in the PHI. Specifically, the PHI gives equal weight to the share of the population living on less than $1.00 a day, the poverty gap (a measure of how far below $1.00 a day the incomes among the poor are), a measure of inequality in income, the prevalence of undernourishment, and the prevalence of low weight-for-age (Gentilini and Webb 2005). One can think

*(continued)*

**Box 8.2. Which Index to Measure Hunger? (*continued*)**

of the three variables measuring poverty and inequality as measures of household access. The undernourishment variable reflects access and national availability while the underweight children variable captures individuals' nutritional outcomes. Compared to the HVI, PHI places more emphasis on access, somewhat less emphasis on nutritional outcomes, and much less emphasis on availability.

An important difference between the HVI and the PHI is the choice of the poverty line. By using the $1.00/day poverty line, the PHI is relevant for measuring progress toward the specific Millennium Development Goals regarding extreme poverty. The HVI index, in contrast, uses the $2.00/day poverty line on the grounds that people living on between $1 and $2 per day can fall quickly into hunger in the face of unstable economic, political, or environmental conditions. Overall, however, the PHI and the HVI tend to agree. The countries for which the HVI and the PHI disagree tend to be those for which the $1.00/day poverty rate is very different from the $2.00/day poverty rate. When these measures are similar, as in Ni-

geria, the indexes give similar results. When they diverge more dramatically, as in Pakistan, the indexes give different impressions.

Despite the above differences, the HVI, PHI, and GHI give similar results in most cases. Differences emerge because each index is measuring a slightly different but related issue. The GHI focuses on the results of recent experience with food deprivation, while the PHI captures extreme poverty, which is a major contributor to hunger. Like the PHI, the HVI captures poverty as a risk factor for hunger. Unlike the PHI, the HVI considers a household's access to food to be at risk if it is poor, though not currently in extreme deprivation. Each of these measures serves a slightly different purpose. The GHI shows where hunger is actually affecting people. The PHI shows where extreme poverty is merging with hunger; and the HVI shows where poverty, food availability, and food deprivation suggest a risk of hunger, even if it is not widely experienced. Since a purpose of this atlas is to map the factors that place people at risk of hunger, it relies on the HVI.

**Table 8.3. Contrasts between the Hunger Vulnerability Index and the Global Hunger Index**

| Country | HVI | GHI | $2/Day poverty | Child growth failure | Child mortality |
|---|---|---|---|---|---|
| *High HVI & low or moderate GHI* | | | | | |
| South Africa | 22 | 7 | 42 | 23 | 7 |
| Kyrgyzstan | 23 | <5 | 51 | 18 | 4 |
| *Very high HVI & serious GHI* | | | | | |
| Uzbekistan | 34 | 11 | 76 | 25 | 4 |
| Lesotho | 36 | 14 | 61 | 45 | 13 |
| Benin | 38 | 15 | 78 | 39 | 15 |
| *Extremely high HVI & serious GHI* | | | | | |
| Nigeria | 42 | 18 | 83 | 43 | 19 |

Source: von Grebmer et al. 2008

Note: Classifications for GHI are: under 5 low, 5–10 moderate, 10–20 serious, 20–30 alarming, over 30 extremely alarming. Classifications in HVI are: under 10 low, 10–19 medium, 20–29 high, 30–39 very high, >40 extremely high.

**Table 8.4. Contrasts between the Poverty and Hunger Index and the Hunger Vulnerability Index**

| | $1.00/Day poverty rate (%) | $2.00/Day poverty rate (%) | PHI | HVI |
|---|---|---|---|---|
| Nigeria | 52 | 83 | Low performance | Extremely high vulnerability |
| Pakistan | 9 | 59 | Medium performance | Very high vulnerability |

Note: PHI rates reported in Gentilini and Webb (2005). Poverty rates for 2005. World Bank, PovcalNet, n.d.

data on availability, poverty, or child growth failure will corrupt the calculated index. Compared to the use of any one component, however, the averaging should diminish the impact of these errors. Third, different

caloric availability thresholds, poverty lines, or anthropometric measures could be chosen and would alter the results. In particular, use of national poverty lines or use of a $1.25/day poverty line would affect both the level of the index and the ranking of countries. For example, use of a $1.25/day poverty line would reduce the poverty rate in Pakistan from 59% to 22%, but would lower the rate in Bolivia only from 29% to 20%. As a result, the index values for the two countries would become much more similar than is reported in table 8.1. In using the higher poverty threshold we are

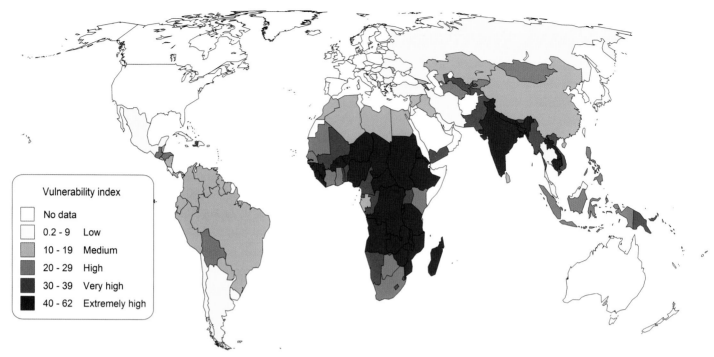

Map 8.1. The Hunger Vulnerability Index. *Note:* This Hunger Vulnerability Index is calculated using the most recent available data on food availability, child growth failure, and $2.00/day poverty as of November 2008. Data are reported in appendix 1.

intentionally trying to capture households that are at risk of lacking access to sufficient food. The $2.00/day threshold may include households with adequate food access, but many (or most) of these households could not tolerate shocks to their income. As discussed in box 8.2, other analyses have used slightly different data to construct alternate indices to measure hunger.

The geography of hunger that emerges in the HVI map (map 8.1) shows the greatest vulnerability levels in Sub-Saharan Africa and South-Central Asia. Southeast Asia shows contrasting levels with Malaysia at the low end (score of 8) and Kampuchea at the extremely high end of the index (score of 40.5). Bolivia (21), Haiti (37), Guatemala (27), and Honduras (23) stand out in Latin America and the Caribbean for having the highest levels of hunger vulnerability. The later sections of this volume consider a number of social, political, and economic conditions that help explain these differences in hunger among countries. First, maps focusing on Uruguay, India, the United States, and Mexico will reveal how hunger manifests itself differently across regions within countries.

# Patterns of Hunger within Countries

Up to this point this atlas has examined the geography of hunger at the country scale. Other spatial scales such as the state, county, municipality, or neighborhood can reveal patterns of hunger within countries. These subnational entities are typically created for administrative purposes such as electing political representatives, collecting taxes, and delimiting school districts. After a while, their boundaries become fixed by laws and inscribed in maps. They eventually function like spatial containers in which programs are implemented. These scales also become the sites of data collection so that administrators (and researchers) can monitor progress in achieving specific goals such as cutting hunger in half by a certain date. The ability to map at these scales depends, of course, on whether data have been collected and are available to researchers.

Country-scale data and maps are useful for country-to-country comparisons. Data are commonly compiled at this scale to present a portrait of a nation in the form of a national average. These national averages are widely used for comparing one country's development progress with others. Country-level data are sometimes grouped into multicountry regions to compare one part of the world with another. Maps 6.2 and 6.3 use this type of regional mapping to contrast child growth failure rates among world regions. However, it is important to examine hunger at other geographic scales to understand its distinctive patterns and the processes producing it.

Maps of hunger vulnerability at the national scale, as in map 8.1, give a general impression of each country's food security situation. For example, map 8.1 suggests that hunger is not a serious problem in Uruguay since its HVI score falls within the lowest vulnerability class. But are there populations in certain regions and neighborhoods within Uruguay whose hunger vulnerability is greater than that suggested by the national average? By mapping hunger at the subnational scale, such as the county in the United States or the department in Uruguay, we obtain a more detailed picture. The following maps reveal that hunger is very unevenly distributed over space in both rich countries and poor ones. Examining hunger at subnational scales is of considerable policy relevance, since it can enable national governments to focus their food security programs in the appropriate locations.

# 9: Child Growth Failure in Uruguay

Uruguay is a small country of some 3.25 million people nestled between Argentina and Brazil. Ninety-two percent of Uruguayans live in urban areas. More than 40% of the country's entire population lives in the capital city of Montevideo located at the mouth of the Rio Plata. What does the geography of hunger look like in a country that is so highly urbanized? Answering this question depends on the availability of data at the relevant scales. Fortunately, Uruguay collects detailed information on child growth failure and makes its findings available to researchers.

The Uruguayan government funds a national food program for primary school children (Programa de Alimentación Primaria or PAP), in which 92% of all students receive some type of nutritional assistance. This can take the form of a simple glass of milk or include three meals a day. The National Administration for Public Education periodically collects data at the departmental scale on child growth to monitor PAM's progress in achieving its objectives. The most recent study was conducted in 2002 and included 56,647 six-year-old children from 1854 schools representing all of the country's 19 departments. The children's heights were measured and compared to the international reference standards for six-year-old boys and girls to identify rates of growth failure (box 6.1). Nutrition experts in Uruguay believe that the growth of well-fed and healthy Uruguayan children conforms to the international "reference child" (ANEP 2003, 13). The Uruguayan data are particularly strong because of the large sample size and the fact that 85% of six-year-olds are enrolled in primary school.

Maps 9.1–9.3 present the results of the Third National Census on the Height of First Grade School Children (ANEP 2003). Unlike the WHO data that reported height-for-age of children under the age of five (map 6.1), the Uruguayan study measured children at the age of six. The national results for growth failure in the two studies differ because they are measuring two different sample populations at different dates.

The 2002 study recorded a national rate of child growth failure of 22.9%. Map 9.1 reveals that this rate varied at the departmental scale from a low of 20.2% to a high of 26.1%. A 6 percentage point difference among departments may not seem significant, but if the departments that have the most serious malnutrition (Duranzo, Salto, San José, Canelones, Tacuarembó, and Cerro Largo) had average rates, hundreds of children would be more healthy. At the very least, this map reveals that not all departments experience the same levels of food insecurity in Uruguay. The departments of Montevideo and Canelones account for 46% of all

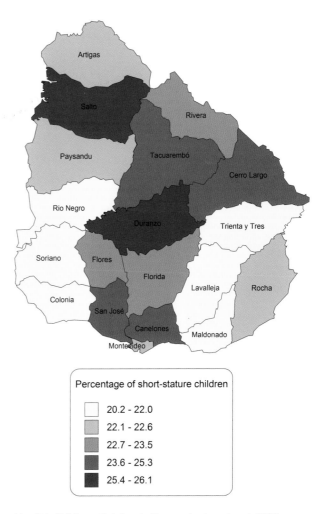

Map 9.1. Child growth failure in Uruguay by department, 2002.

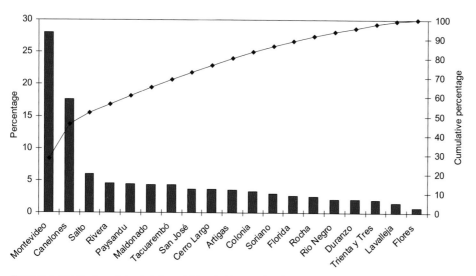

Figure 9.1. Distribution of short-stature children in Uruguay by department, 2002. *Source:* ANEP 2003.

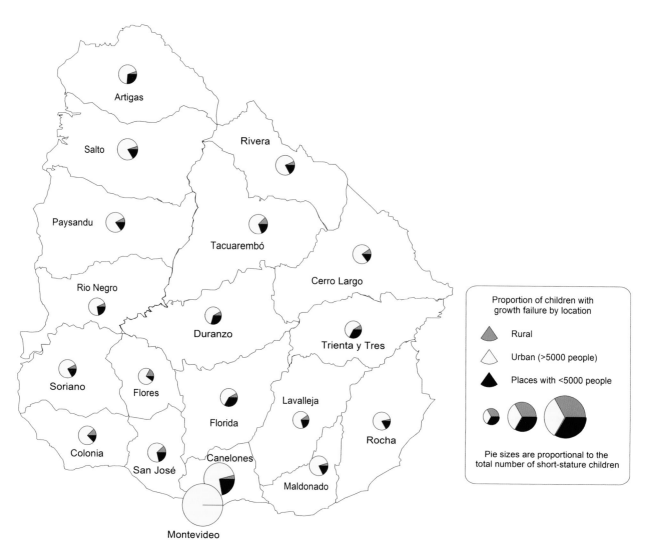

Map 9.2. Rural-urban differences in child growth failure in Uruguay, 2002.

Map 9.3. Child growth failure in Montevideo by neighborhood, 2002.

growth failure measured in the 2002 survey (figure 9.1). This finding alone can help policy makers in Uruguay target their food and nutrition activities to have the greatest possible impact.

Map 9.2 reveals that there are important differences in hunger vulnerability *within* departments, particularly between rural and urban areas. Children residing in urban areas, defined as communities with populations of greater than 5000 people, accounted for 81% of the total number of short-stature children. Just 5% of the total number of these children resided in rural areas, while 14% lived in communities with fewer than 5000 people. Since only 8% of the population is nonurban, most children suffering undernutrition are urban, even though rates of growth failure in rural areas and small towns are high (about 40%). The urban concentration of food insecurity suggests that nutritional interventions in cities of greater than 5000 people would have a large impact. Yet, not all urban children suffer

the same levels of food insecurity. In the Department of Canelones, where there are some twenty cities with over 5000 people, there was a three-fold difference in growth failure rates between the cities of Lamas y Médanos de Salymar (12.1%) and Villa Crespo y San Andrés (36%). These findings can give further insight into how the government of Uruguay might target PAP's activities.

Hunger vulnerability also differs within urban areas. Taking the case of Montevideo (map 9.3), we find striking differences in rates of child growth failure among neighborhoods. These range from 9.5% in the affluent neighborhood of Punta Gorda to 37% in the low-income neighborhood of Tres Ombúes. In general, the pattern of child growth failure is lowest in the central city neighborhoods (with the exception of Villa Munoz) and is highest in the northwestern barrios and in the outer-ring neighborhoods of Tres Ombúes, Jardines del Hipódromo, and Casavalle. One out of three children in these poor neighborhoods showed serious signs of growth failure in the 2002 study.

In summary, maps 9.1–9.3 and figure 9.1 reveal the advantages of examining the geography of hunger at multiple scales. We find that national averages mask considerable differences among regions and between and within urban areas. It is only at the intraurban scale, for example, that we discover child growth failure rates in some neighborhoods to be 10–15 percentage points higher than the national average. The departmental-level maps and graph point to areas of concentrated undernutrition. Exposing these spatial differences is important for directing health and nutrition pro-grams to areas where needs are greatest. Uruguay's primary-school food program appears to be reaching those children who most need nutritional assistance. Three-quarters of the children surveyed participate in the school lunch program. Those who do not partici-pate are children showing "normal" growth patterns. The multiscale geographical data and maps are useful tools for PAP in tracking the hunger problem across the country. They confirm that governments should make it a priority to collect subnational data in order to bet-ter unmask the face of hunger that is too often hidden behind country-level figures.

# 10: Food Insecurity in the United States

To measure hunger vulnerability in the United States, the Department of Agriculture administers regular surveys in which people are asked questions to determine whether or not they are food secure. In these surveys, "food security" means that over the past 12 months households have obtained sufficient food to lead healthy and active lives. Until 2006, a household would be classified as "food insecure without hunger" if respondents indicated that they ran out of food before they had money to buy more, couldn't afford balanced meals, and worried that they would run out of food. If such conditions existed and respondents reported that adults ate less than they felt they should and also skipped meals in three or more months during the previous 12 months, the household would be classified as "food insecure with hunger." In 2006 the USDA altered its language to avoid using the word "hunger" (MacKinnon 2006). We maintain the original labels here.

Map 10.1 indicates the distribution of food-insecure households in the United States. The map includes households that were classified as food-insecure with hunger and those that were food-insecure without hunger. On average 11.3% of US households were food-insecure with or without hunger in the 2004–6 period. This was up from 11% in 2001–3. As table 10.1 shows, there was a peak in overall food insecurity at 11.9% in 2004. Food insecurity with hunger continued to rise after that point. The map shows that food insecurity was not evenly distributed across the country. Texas, Mississippi, and New Mexico recorded the highest rates of hunger risk, with over 15% of their households food-insecure, while six states (Delaware, Hawaii, New Hampshire, New Jersey, North Dakota, and Virginia) reported food insecurity rates of less than 8%.

There is a distinct regional pattern to food insecurity in the United States. With the exception of Florida, hunger rates were above the 11.3% national average in the South, below average in the Northeast with the exceptions of Maine and Rhode Island, and around the national average in the West and Midwest. Map 10.2 gives

a more dynamic view of hunger in the United States. The map shows statistically significant declines and increases in hunger rates between 2001–3 and 2004–6. Hunger rates fell in four states (California, Florida, Hawaii, and Montana) but increased at statistically significant rates in 13 others plus the District of Columbia.

Table 10.2 shows that hunger vulnerability is also distributed unevenly in the United States based on race and on income and that families with children are more likely to suffer food insecurity than those without them.

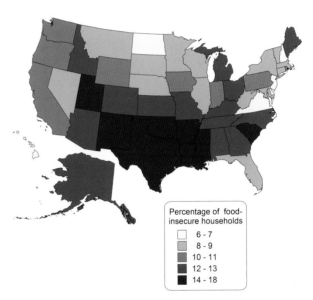

Map 10.1. Food insecurity in the United States, 2004–6.

**Table 10.1. Trends in food insecurity in the United States, 1999–2006 (% of households)**

|  | Food-insecure (with or without hunger) | Food-insecure (with hunger) |
|---|---|---|
| 1999 | 10.1 | 3.0 |
| 2000 | 10.5 | 3.1 |
| 2001 | 10.7 | 3.3 |
| 2002 | 11.1 | 3.5 |
| 2003 | 11.2 | 3.5 |
| 2004 | 11.9 | 3.9 |
| 2005 | 11.0 | 3.9 |
| 2006 | 10.9 | 4.0 |

Source: Nord, Andrews, and Carlson 2007.

In 2006, 21.8% of all (non-Hispanic) black households were identified as food-insecure. In comparison, 7.8% of (non-Hispanic) white households experienced food insecurity. As in the case of Uruguay, this subnational data reveal that national averages often hide important variations in hunger vulnerability within a country.

As in other countries, hunger and poverty are closely related within the United States. Map 10.3 presents the share of the population in each US county whose in-

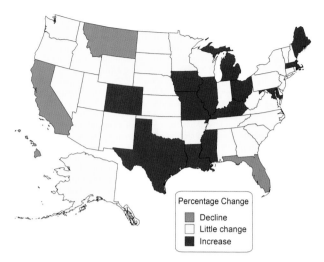

Map 10.2. Changes in hunger prevalence in US households, 2001–3 to 2004–6.

**Table 10.2. Distribution of food insecurity in the United States, 2006 (% of households)**

| Household category | Food-insecure (with or without hunger) | Food-insecure (with hunger) |
|---|---|---|
| All households | 10.9 | 4.0 |
| Households with children | 15.6 | 4.3 |
| White (non-Hispanic) | 7.8 | 3.1 |
| Black (non-Hispanic) | 21.8 | 8.0 |
| Hispanic | 19.5 | 5.7 |
| In poverty | 36.3 | 14.8 |

Source: Nord, Andrews, and Carlson 2007, 10.

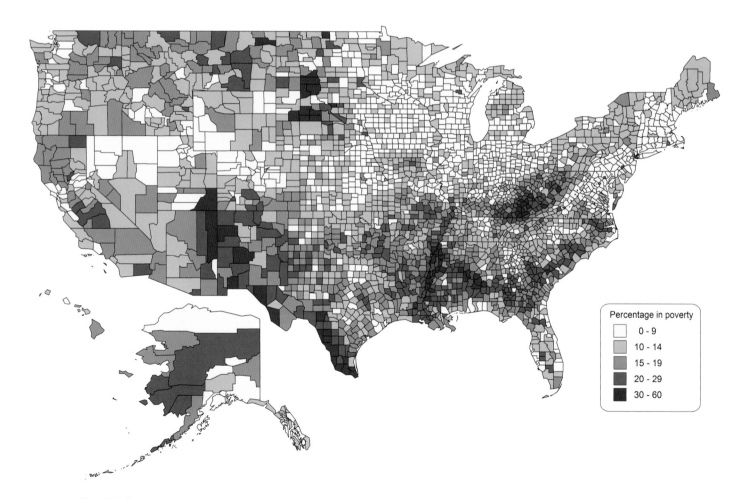

Map 10.3. Poverty per capita in the United States, 1999.

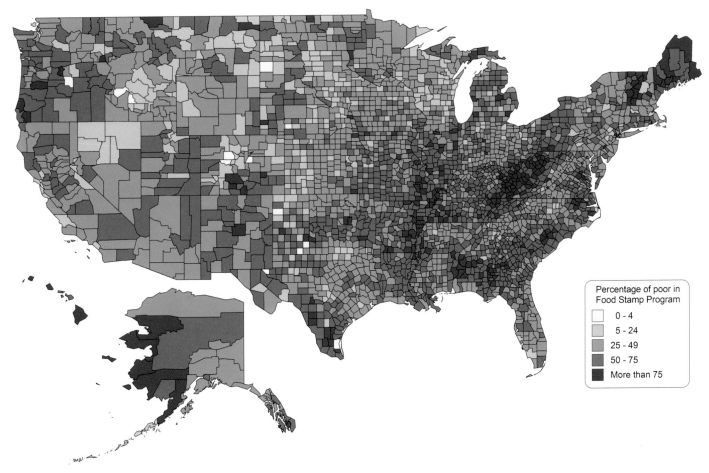

Map 10.4. Percentage of poor households participating in Food Stamp Program, 1999.

come placed them below the poverty line. In this case, a poverty line was established based on the costs of purchasing the basic requirements for food and shelter in the area in which a household lived. The average poverty rate for the country is about 14%, but county-level poverty rates ranged from under 5% to over 50%. Given this kind of variation, a national average will misrepresent conditions in most of the country. As can be seen from the map, poverty, like hunger, is concentrated in the South and Southwest. Poverty rates are below average in the Midwest and East Coast. Relatively few counties have poverty rates that are close to the national average.

The county-level poverty map shows great variation even within states. Texas, for example, has high rates of food insecurity, but poverty is concentrated in counties along the border with Mexico. Poverty rates in northeastern Texas are very low. West Virginia had relatively low food insecurity based on the state-wide data (map 10.1), but the county-level poverty map shows high poverty in the southwest of the state and low poverty in the Northeast. In both these cases, the state average can misrepresent reality.

The poor need not suffer hunger or food insecurity if social safety nets are in place to support them. The USDA's Food Stamp Program exists for this purpose. Through this program, people whose incomes place them at risk of hunger can acquire coupons or debit cards to purchase food. In principle, if every poor household had access to and used a sufficient amount of food stamps, no one in the US would be food-insecure. However, not all eligible Americans use food stamps. Map 10.4 shows the number of people receiving food stamps as a share of the population in poverty (most of whom are eligible for food stamps). County by county, participation rates range from under 25% to over 75% of eligible households. In general, participation rates are higher in those counties where poverty rates are high. Yet some states seem to have different rates of success in reaching their poor. West

Figure 10.1. Lunchtime at a soup kitchen in east-central Illinois. The 2008 economic recession led to a 20% increase in demand for food at the nation's food banks, food pantries, soup kitchens, and emergency feeding centers (Feeding America 2008).

Virginia, for example, reports a high share of its poor participating in the Food Stamp Program, especially in the high-poverty counties. In contrast, none of the border counties in Texas have participation rates over 75%. Given the high rates of poverty in these counties, low food stamp participation per poor person suggests that many people are falling through this food security safety net and are experiencing hunger.

Some food-insecure households obtain food from emergency food providers such as food pantries, soup kitchens, and emergency shelters (figure 10.1). Food pantries give food to people to prepare in their homes. Soup kitchens and emergency shelters prepare food that is consumed at their locations. Feeding America (formerly named America's Second Harvest), a network of emergency food providers, served an estimated 25 million different individuals in 2005 (Cohen, Kim, and Ohls 2006). Seventy percent of the households served by the network identified themselves as food-insecure. A third of these households had experienced hunger as defined by the USDA indicators of food insecurity with hunger.

Many of the individuals participating in these private-sector food relief programs also receive public assistance. Thirty-five percent of the households surveyed by Feeding America received Food Stamp Program benefits, although many more were eligible. More than half of the households with children between the ages of one and three years were enrolled in the government-run Supplemental Nutrition Program for Women, Infants, and Children (WIC). Sixty-two percent of the food-insecure households with school-aged children participated in the federal school lunch program, while 51% benefited from the school breakfast program. It is this combination of public and private programs that provide a safety net to the poor and food-insecure households in the United States.

# 11: Malnutrition in India and Mexico

## ANEMIA IN INDIA

Severe anemia is a significant nutritional problem in India that is unevenly distributed within the country. In 2005–6, 2.9% of Indian children less than five years of age suffered severe anemia, putting them at an immediate health risk and reducing their capacity to learn and grow. Map 11.1 shows that 8 out of India's 29 states had rates above the national average. The states of Punjab and Rajasthan stand out with the highest prevalence, with rates that were more than double the national average. Boys and girls are equally afflicted, but rates are much higher in rural areas (IIPS and Macro International 2007, 287–90). These findings can help public health workers focus their programs in those states and areas where intervention is most urgent. However, nutritional problems in India are not all distributed in the

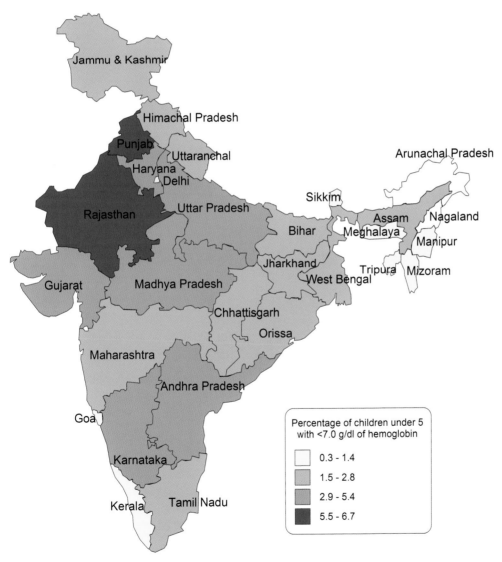

Map 11.1. Severe anemia in India, 2005–6.

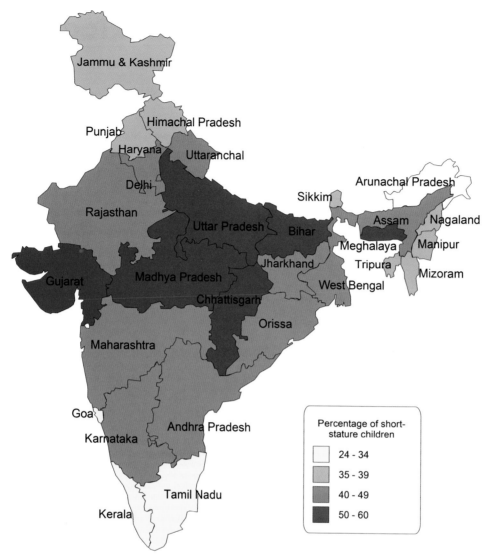

Map 11.2. Child growth failure in India, 2005–6.

same way. The pattern of growth failure (map 11.2) is distinct from that of anemia.

Map 11.2 shows the percentage of short-stature children, generally a sign of undernourishment. Comparison of the distribution of anemia and child growth failure shows that anemia is present where food intake is generally low. The north central states in India have both high rates of short-stature children and high prevalence of anemia. However, the maps also show that some regions with the most intense problems with anemia have relatively lower rates of child growth failure (e.g., Punjab and Rajasthan). These discrepancies can partly be explained by the findings that anemia was most common in low-income households, and in those in which the mother had no formal education (IIPS and Macro International 2007, 287–90). Availability of fish in the diet results in low prevalence in the far south of India, while a combination of poverty and vegetarianism might intensify iron-deficiency anemia in Rajasthan. Anemia in Punjab and Rajasthan may also result from disease and sanitation problems inhibiting the body's ability to absorb iron in the diet.

## OBESITY IN MEXICO

Maps 5.1 and 5.2 highlighted the growing problem of obesity even in countries where hunger is common.

Map 11.3. Adult obesity in Mexico, 2000.

**Table 11.1. Trends in obesity among Mexican women aged 20–49 years (%)**

| Year | 1988 | 2000 |
| --- | --- | --- |
| National rate | 9.4 | 29.0 |
| Urban | 9.6 | 30.7 |
| Rural | 9.1 | 27.2 |
| Socioeconomic status | | |
| Low | 9.7 | 25.1 |
| Medium | 10.6 | 32.8 |
| High | 8.4 | 29.3 |

Source: Barquera et al. 2006

**Table 11.2. Rates of diabetes mellitus and hypertension among adults in Mexico, 1994–2000 (%)**

| | Diabetes mellitus | | Hypertension | |
| --- | --- | --- | --- | --- |
| | 1994 | 2000 | 1994 | 2000 |
| Gender | | | | |
| Female | 3.7 | 8.3 | 27.8 | 30.9 |
| Male | 4.3 | 7.6 | 39.3 | 39.2 |
| Location | | | | |
| Rural | — | 7.2 | — | 33.0 |
| Urban | 4.0 | 8.8 | 26.1 | 34.2 |
| Region | | | | |
| North | 3.4 | 9.8 | 25.4 | 38.4 |
| Central | 5.0 | 7.6 | 30.0 | 33.7 |
| South | 5.2 | 7.3 | 23.6 | 30.4 |
| Mexico City | 1.7 | 8.9 | 23.7 | 27.9 |

Source: Barquera et al. 2006.

The case of Mexico is revealing (table 11.1). Women's obesity increased by a factor of three between 1988 and 2000, rising from 9.4% to 29% (Barquera et al. 2006). Table 11.1 indicates that obesity rates are high among women of all socioeconomic classes in both urban and rural areas, although urban rates tend to be higher.

Socioeconomic status remains important in explaining the regional geography of obesity. A larger percentage of the adult population is obese in the higher-income northern and central regions of Mexico in comparison to the poorer southern region (map 11.3).

These obesity trends are worrisome because of their implications for chronic diseases. A dramatic rise in diabetes and hypertension is linked to the prevalence of obesity in all regions (table 11.2).

# Part II: The Sources of Hunger

In part I we focused on the most common indicators of hunger and then developed an alternative measure, the Hunger Vulnerability Index, which reflects our efforts to better locate hunger in the world. Part II examines the relationship between hunger vulnerability (map 8.1) and the many factors that may explain the patterns that appear in this map. The factors that we examine derive from the conceptual framework ("The Sources of Hunger") presented in the introduction to this work (figure 1.2).

The structure of part II thus follows that of our conceptual framework. Each of the major boxes in figure 1.2 (resources, technology, institutions and power relations, poverty, exacerbating conditions and events) becomes a subsection in the following pages. In each section we ask "How important is a country's resource base (or technology or institutions and power relations) to the pattern of hunger vulnerability displayed in map 8.1?" We also examine the dynamic relationships among these factors to explain the geography of hunger. We argue, however, that one cannot explain hunger vulnerability by simply adding up in a checklist-like fashion the many factors that appear to influence food availability, household food security, and individual nutrition. The recurring message of part II is that hunger is always linked to poverty and social vulnerability. These realties are ultimately rooted in policies and political economies that fail to protect the poor from exacerbating conditions and events that push people deeper into poverty, heighten their vulnerability, and produce hunger.

# National Resources

How important is a country's resource base to its capacity to meet national food availability needs, achieve household food security, and provide good nutrition to individuals? In our model illustrating the causes of hunger, the resources available in a country play a central role. The model takes a broad view of resources. It considers natural resources like land, water, and minerals; human resources like education and health; and human-made resources like roads, utility grids, and sanitation systems. Our model also suggests that technology and the institutions and power relations that govern access to resources influence what a country generates from its resource base and who enjoys the benefits. The following maps and graphs attempt to plot the distribution of resources globally and to relate that distribution to patterns of hunger vulnerability.

In general, the maps do not show strong correlation between the resource wealth of a country and its hunger vulnerability. We find many cases where resource abundance does not diminish hunger and where resource scarcity exists with food security. These outcomes emerge because many other factors relating to political power, economic distribution, and technology influence the effect of resources on hunger. Human and natural resources are extremely important to our well-being. But whether a given country's resources lead to hunger vulnerability or not often depends on their management and distribution.

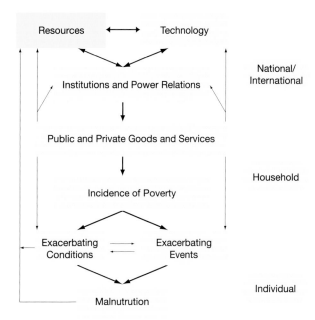

# 12: Population Growth

*The power of population is so superior to the power of the earth to produce subsistence for man, that premature death must in some shape or other visit the human race. The vices of mankind are active and able ministers of depopulation. They are the precursors in the great army of destruction; and often finish the dreadful work themselves. But should they fail in this war of extermination, sickly seasons, epidemics, pestilence, and plague, advance in terrific array, and sweep off their thousands and tens of thousands. Should success be still incomplete, gigantic inevitable famine stalks in the rear, and with one mighty blow levels the population with the food of the world.*

**Thomas Malthus, 1798**

Over 200 years ago, the scholar Thomas Malthus argued that growth in human population inevitably places so much pressure on the earth's resources that "gigantic" famines become unavoidable. He believed that the demand for food by more and more people would exceed supply ("the power of the earth to produce subsistence") so that hunger and death would inevitably follow. Malthus tried to quantify this relationship by arguing that populations grow at a geometric rate (2, 4, 8, 16, 32) while food production increases at a slower arithmetic rate (2, 3, 4, 5, 6). Today Malthus's prediction has become so widely accepted that many people assume when hunger and famine occur that "overpopulation" is the principal cause. Today governments give emergency food aid to people in famine situations and promote population control (e.g., family planning) as a long-term solution. We cling, like Malthus, to the notion that hunger stems from population pressure on resources. This view is misleading because it exaggerates one dimension of a multifaceted problem and diverts attention from the political and economic sources of hunger.

The data on food production do not support the Malthusian view. It is true that population has grown and hunger has persisted, but as figure 12.1 shows, food production per person has risen during the last 50 years.

The most dramatic increases have been in Asia and Latin America. Food production per capita did fall in Africa during the 1970s and early 1980s, but it has been on an upward trend since then. Since 1990, the region with the most pronounced decline in food production per capita is Europe, where famine is not an issue and obesity is a more serious nutritional problem than hunger. The record refutes Malthus. Thanks largely to improvements in technology, the "power of the earth to produce subsistence" has been superior to the power of population to grow. Nonetheless, the world continues to experience recurrent famines and even when famine is avoided, people suffer chronic hunger.

It is a myth that "overpopulation" breeds hunger, but myths often contain elements of truth. It is true that the doubling of a country's population in a short period of time can place pressure on existing resources. But the distribution of these resources and inequalities in income, entitlement, and opportunity are more important in generating hunger than overpopulation. It is important to ask how many people seek a share of some resource. It is equally important to ask how that resource is being shared, and what is being done to increase its supply or its productivity. How the food problem is framed is critical to how it is analyzed. Too often a focus on the real pressure that population growth can place on natural resources distracts attention from analysis of how to reallocate wealth, income, or opportunity to alleviate poverty and hunger.

In our view, politics and technology interact with available resources to determine hunger outcomes (figure 1.2). A rising population will not produce famine if new technologies allow food production or incomes to rise in pace with population. By the same token, increasing food production per capita may not reduce hunger if food and incomes are distributed unequally. Because both investment in new technology and the distribution of income are affected by policy, there is nothing inevitable in the relationship between population and hunger.

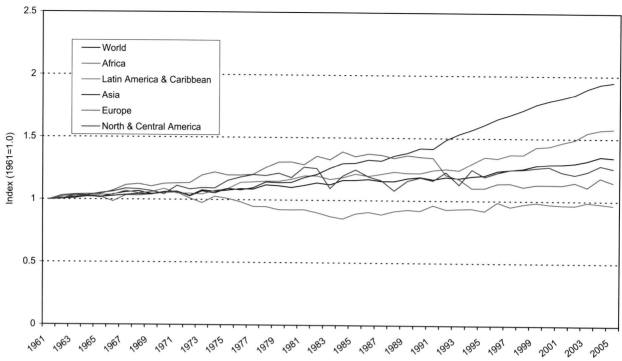

Figure 12.1. Food production per capita in the world. *Source:* FAOSTAT 2007a.

To assess whether population growth explains hunger in a specific country, one must first ask what resource might be under pressure from population. If it is land, then we must examine land distribution within that country. If it is water, then we need to determine who has access to safe water and who does not. It will not take long to see that landholding patterns and the distribution of clean water and sanitation are highly charged political issues in many countries. Where the political system in a country excludes the poor or minority groups from access to land, technology, education, or opportunity in general, then population growth among those groups may indeed be impoverishing. At the same time, poor people who are excluded from many opportunities may perceive that having large families with many potential income earners is a sensible choice to make. Political and economic exclusion can also leave these populations with little access to services that offer family planning. In summary, an emphasis on population growth often draws attention away from many factors that governments can influence to enable their entire population to better manage national resources like land and water.

A first glance at the maps does show some corre-spondence between high population growth and high rates of hunger vulnerability (cf. maps 8.1 and 12.1). Indeed, the scatter diagram in figure 12.2 shows a trend of increasing hunger vulnerability with higher population growth rates. However, the wide dispersion of the data in figure 12.2 means that there are many exceptions to this pattern. The maps show that many African countries have high hunger vulnerability and high population growth rates, but the Arabian Peninsula and Middle East have extremely high population growth (map 12.1), with very low hunger rates (map 8.1). India and Mexico have similar high rates of population growth in map 12.1, but very different rates of hunger vulnerability. In all these cases, poverty rates are largely responsible for differences in hunger outcomes. With relatively high national incomes and reasonably effective social safety nets (like food stamps for the poor), many countries can have rapid population growth with low hunger. Meanwhile, one of the countries with the most severe hunger problems in Africa is Zimbabwe. As the population growth map shows, Zimbabwe has a very low population growth rate. This low rate of growth is not due to low fertility rates, as is the case in Europe. Rather, it is a result of high mortality from

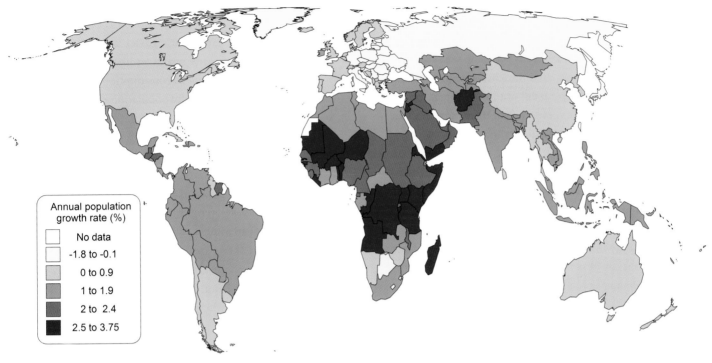

Map 12.1. Population growth, 2006.

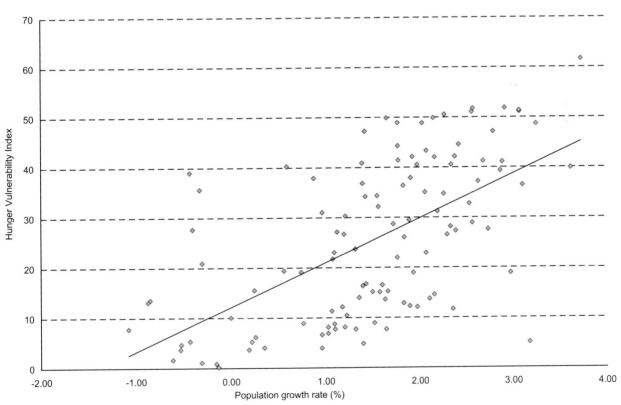

Figure 12.2. Population growth and hunger vulnerability. *Sources:* World Bank 2008a; WHO 2008c; FAO 2008b; World Bank, PovcalNet, n.d.

**Box 12.1 Population Pressure and Consumption Pressure**

The idea that population places pressure on the earth's resources sometimes leads people to see rapid population growth as a global environmental threat. Many aspects of our natural environment are under threat, but it is people's consumption rather than their numbers that places stress on the earth's capacity to provide.

Global climate change due to the emission of greenhouse gasses represents one way in which the natural environment is under threat from high consumption. Based on their consumption patterns, citizens of the United States generate an average of 20 metric tons of carbon dioxide ($CO_2$) emissions per year. In contrast Kenyans are responsible for only 0.25 metric tons of $CO_2$ emissions annually per person. By this measure, each additional American generates as much greenhouse gas as 80 Kenyans.

High emission of greenhouse gasses is matched with high consumption of many resources like minerals, timber, and petroleum. As long as consumption patterns are unchanged, slow growth in the population of the United States and other wealthy countries will place far greater strain on the earth's resources than more rapid population growth in Kenya and other developing regions (see figure 12.3).

It is true that population pressure can make living more difficult in developing countries like Kenya. As map 13.2 suggests, decreasing availability of land per person in places where there are limited nonfarming options requires people and governments to find new means of securing livelihoods. But on a global scale it is the rich of the world and not the poor who threaten the world's resource base.

HIV/AIDS. In the unfortunate case of Zimbabwe and other countries where AIDS has become epidemic, mortality and reduced population growth are contributing to a hunger crisis (see map 29.1).

Thomas Malthus's dire predictions that population growth would inevitably lead to hunger were mistaken. Yet the notion that hunger is a result of overpopulation remains popular. Unfortunately, a focus on population as the cause of hunger problems is rarely useful and frequently leads to a conclusion that the hungry will remain deprived until they control their own fertility. Blaming the victims of hunger for their condition in this way ignores the factors that cause poverty in the first place. The poor are hungry because of their poverty and their poverty is rooted in a mix of political, economic, and physical conditions. Experience has shown that the same fundamental changes that reduce poverty also reduce rates of population growth and hunger. Around the world, as men and women have found new opportunities for education and good em-

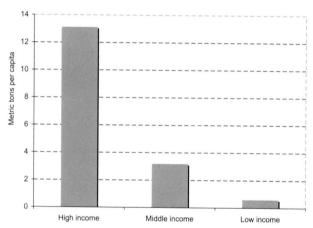

Figure 12.3. Per capita carbon dioxide emissions for countries by economic group. *Source:* World Bank, *World Development Indicators Online,* n.d.

ployment for themselves and their children, poverty, hunger, and population growth rates have all fallen. Population decline in the absence of such new opportunities has rarely occurred without coercion and has rarely signaled improvements in people's well-being.

# 13: Arable Land per Capita

Is there a link between vulnerability to hunger and the amount of potentially farmable (arable) land in a country? It is tempting to think that shortages of arable land per capita and hunger must be related. The maps reveal no one-to-one correspondence. In some cases, there does seem to be a relationship. Many Central and Southern African countries have relatively low amounts of arable land and score high on the Hunger Vulnerability Index (map 13.1 and map 8.1). However, counterexamples immediately stand out. Sudan, the Central African Republic, and Zambia have relatively high quantities of arable land per capita but also high rates of hunger vulnerability. In South America, Venezuela, Colombia, Ecuador, Peru, and Chile have relatively low amounts of arable land per capita but also have low hunger vulnerabilities. The amount of arable land in a country matters, but not in isolation from other conditions that affect food insecurity. These other factors include the utilization of land and the availability of nonagricultural employment.

The utilization of arable land may help explain the discrepancies between arable land per capita and hunger vulnerability. Good farmland that in theory could employ people as either farmers or farm laborers is often underused. Underutilization could emerge when a small number of people control vast amounts of land or when small-scale farmers lack the skills, tools, or inputs to use their land. For example, a study conducted following Zimbabwe's independence in 1980 focused on land utilization among large-scale commercial farmers in regions where most of the country's prime agricultural land is located (Weiner et al. 1985). Twenty-six hundred farmers controlled three-quarters of the region's farmland. Their average farm size came to 1640 hectares. But, on average, just 168 hectares were actually under crops on these farms. The authors

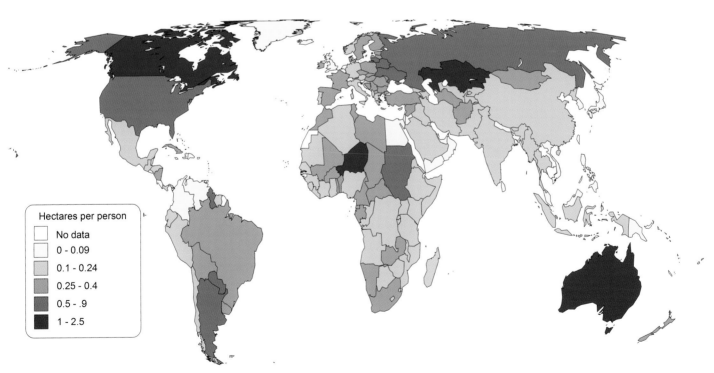

Hectares per person

| | |
|---|---|
| | No data |
| | 0 – 0.09 |
| | 0.1 – 0.24 |
| | 0.25 – 0.4 |
| | 0.5 – .9 |
| | 1 – 2.5 |

Map 13.1. Arable land per person, 2005.

**Box 13.1. The Landless Workers Movement of Brazil (O Movimento dos Trabalhadores Rurais Sem-Terra)**

The amount of arable land per capita in a country is not a good predictor of hunger vulnerability. In some societies, abundant land coexists with widespread landlessness. This is particularly true in Latin American countries in which a small percentage of the population controls a large amount of land. Brazil is notorious for its pattern of unequal landholdings, where 4% of landowners control more than 50% of the nation's agricultural land. In addition to inequitable landholdings, a small percentage of the population controls a large share of national income. In 2003, the richest 20% of Brazil's population controlled 62% of total income. The share of the poorest 20% of the population amounted to just 3%. Almost a fifth of its population lives on less than $2.00 a day. This volatile mix of poverty and inequality produces hunger, despair, and class conflicts. It has also been the source of a highly successful social movement known as Sem-Terra that has obtained agricultural land for landless farmers and farmworkers through peaceful protests and land occupations. Since its official beginning in 1984, Sem-Terra has settled 350,000 families on some 10 million hectares of agricultural land in 22 of Brazil's 26 states. One of the movement's objectives is to obtain obtain "land for those who work it." This goal

of making idle land productive is condoned by the 1988 Brazilian constitution; Article 186 allows the federal government to expropriate and redistribute land that is not meeting its "social responsibility to be productive." The government's response to squatter groups has vacillated from legalizing their land occupations to imprisoning their leaders for illegal land invasions. The movement's activities have been spurred by the democratic opening following the end of military rule in 1985, by the institutional strength and progressive leadership of certain religious organizations, and landlessness that has made land occupations necessary to survival. Gaining access to land is necessary for improving rural livelihoods, but it is insufficient. The most successful Sem-Terra settlements are those that have pressured the federal government to implement agrarian reform policies that improve smallholder access to agricultural credit and markets, housing, and educational and health programs. Such reforms are critical for reducing the social inequities and poverty that generate hunger and social conflicts in a country where there is ample arable land per capita (Wright and Wolford 2003; Wolford 2004).

concluded that depending on the district, between two-thirds and one-half of Zimbabwe's prime agricultural land was not being farmed in 1981–82. In addition, high levels of mechanization on large farms contributed to high rates of unemployment in rural areas (Weiner et al. 1985, 256–76).

Zimbabwe's land reform policies in the 1980s targeted this prime but underutilized agricultural land. A land redistribution program provided both land and agricultural extension services to landless farm laborers, families farming marginal soils in less well-endowed regions, and politically well-connected individuals. The result of this redistribution was that more land came into production and more people were employed and fed. Less orderly land redistribution in Zimbabwe in 2001–3 had the opposite effect. In that case, large-scale farmers who were productively using land were forcibly removed and replaced by inexperienced and ill-equipped settlers who lacked support,

skills, and inputs. Land utilization fell, contributing to declining agricultural output and rising poverty and hunger.

Large amounts of arable land might exist in a country, but inequalities in landholdings can lead to a situation where a minority of the population controls a disproportionate share of the available land. There might be adequate amounts of agricultural land available but high rates of landlessness contribute to widespread hunger. This is the case in Brazil, where land is abundant but millions go hungry. Relative land scarcity is political and economic in origin, and demands a political-economic solution. The Sem-Terra movement in Brazil is a grassroots social movement that seeks to redistribute underutilized land to the landless as a means of addressing poverty and hunger (see box 13.1).

Discussions of access to arable land and food security must also consider other land users besides farmers. These include pastoral peoples who raise livestock

**Box 13.2. When Does Arable Land per Capita Relate to Hunger Vulnerability?**

As the discussion of map 13.1 indicates, there is very little relationship between arable land per capita and hunger when looking at all countries. This point is apparent in the top panel of figure 13.1, which plots the land per capita and hunger vulnerability scores for all the countries in map 13.1. There is little pattern to the data here. The countries with more land per capita rarely have low hunger vulnerability, but countries with little land per person have a wide range in hunger conditions. The absence of a relationship is partly because in many countries people do not rely on land to make their living.

The second panel in this figure plots land per capita and hunger vulnerability scores for those countries in which over half the population lives in rural areas. For this subset of countries there is a tendency for more land to be associated with lower hunger, but the relationship is not as strong as one might expect. Two factors contribute to a situation where more land does not necessarily imply less hunger. One is that not all land is of equal quality. Some of the countries with relatively abundant land and relatively high hunger vulnerability are in semiarid regions where crop yields are low and unreliable. Moreover, as the text stresses, availability of land does not ensure access to land or income from it. In some cases, relatively abundant land per capita is found with widespread landlessness, poverty, and hunger.

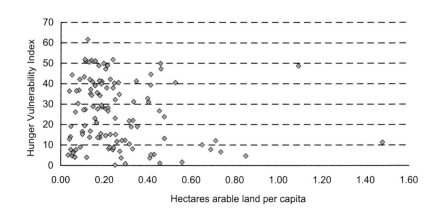

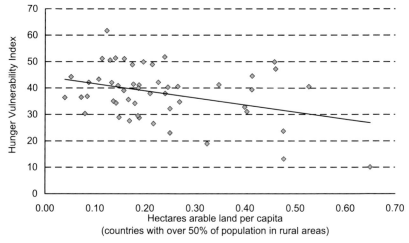

Figure 13.1. Hunger vulnerability and arable land per capita. *Sources:* World Bank 2008a; WHO 2008c; FAO 2008b; World Bank, PovcalNet, n.d.

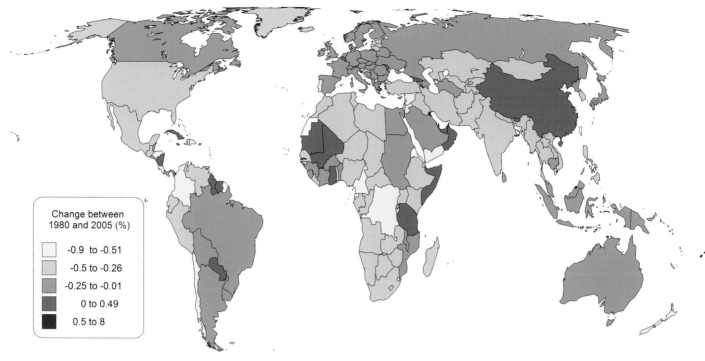

Map 13.2. Change in arable land per person, 1980–2005.

for subsistence and the market, and groups that rely upon access to forest resources such as rubber and nuts. These peoples' livelihoods depend on access to rangelands and forests rather than farmland. Moreover, factory workers, artisans, traders, and many other people earn their daily bread (or rice or porridge) without using land at all. Access to arable land has little direct effect on hunger for these populations.

Examination of arable land per capita indicates whether the natural resource base of a country is capable of feeding its population. However, it raises too many questions about land use, land access, and land productivity to serve as a good indicator of hunger in the world. There are too many intervening variables such as access to land, labor and capital, employment, and market conditions that make it extremely difficult to use availability of farmland to explain hunger. The examples of small countries such as Japan, Singapore, and the Netherlands with large, well-nourished populations reinforce the argument that there are larger political and economic processes that determine a population's vulnerability to hunger (box 13.2).

Another way to examine the relationship between availability of land and hunger is to look at trends in arable land per capita and hunger vulnerability. Map 13.2

shows changes in the amount of arable land per capita between 1980 and 2005. In most cases, the amount of land suitable for agriculture has not increased at the same rate as population growth. Nevertheless, there are some countries (Nicaragua, Suriname, Paraguay, Mauritania, Mali, Ghana, and China) that have experienced growth in arable land over this period. How is this possible? Land that was formerly considered unsuitable for cultivation (mangrove swamps, semiarid areas) has become arable due to the development of irrigation infrastructure or the deforestation of mangroves and other tropical forests.

Has increased or decreased farmland per capita affected a population's vulnerability to hunger? The maps reveal no apparent trend. Countries that have experienced increases in arable land show very different rates of hunger vulnerability, but generally declining rates of undernourishment (map 3.2). Meanwhile, countries such as Libya, Tunisia, Syria, Jordan, and Chile that experienced declining farmland per capita showed relatively low hunger vulnerability and declining undernourishment. The trends in land availability and hunger vulnerability correspond to each other only when conditions like poverty and reliance on farming for income are prevalent.

Figure 13.2. Mural showing famers plowing a field with oxen painted on the inside wall of the village library of Koumbia in southwestern Burkina Faso.

Map 13.2 reveals declining arable land per capita in the United States and in East Africa. This trend does not negatively affect food security in the United States, where there are many other income-earning opportunities outside of agriculture. But in countries like Kenya and Tanzania, where the majority of the population depends on farming and livestock raising, decline in land per capita may contribute to hunger and poverty because political, economic, and technological conditions limit jobs in the nonagricultural sectors of these countries. Contrasting livelihood options between the United States and East Africa suggest that focusing only on arable land per capita distracts our attention from these larger political-economic conditions that are always central to our assessments of hunger vulnerability. Declining availability of farmland per person is most likely to contribute to hunger in places where the arable land per capita is already in short supply and large shares of the population rely on the land for their livelihoods. In such settings it is critical that public policy contribute to effective use of land resources while enabling people to earn incomes independent of holding farmland.

# 14: Environmental Systems Health

Arable land is just one natural resource. Consideration of additional resources might yield more insight into the relationship between natural resources and hunger. The notion of "environmental system" is helpful in this regard. In an attempt to create an index of environmental sustainability, researchers at Yale and Columbia Universities evaluated the health of different countries' environmental systems based on air quality, biodiversity, land availability and quality, and water quality and availability. On the basis of this assessment, they created an environmental system index that ranges from 0 to 100. High scores indicate that the environment is in good health; low scores suggest poor health (Esty et al. 2005). This environmental systems component of the larger Environmental Sustainability Index can be a gauge of the general condition of a country's natural resources.

Map 14.1 shows that the quality of environmental systems varies considerably across world regions. Regions with low population densities and low levels of industrialization have higher scores. Such areas tend to have low levels of air and water pollution and a relative abundance of nondegraded land. For countries that are small, densely populated, and industrialized such as the Netherlands and Belgium, we see very low measures of environmental health. Large, agricultural-based countries like Mongolia tend to score high on the environmental systems index, as do countries with large areas of undisturbed land like Canada and Australia. Since agricultural economies tend to have high rates of poverty, they often experience considerable hunger despite the sound health of their environmental systems as measured here. At the same time, high-income countries like Belgium and Japan have less of a hunger problem, even if their natural resource base is very limited. As a result, the map shows no strong relationship between the quality and amount of a country's natural resource stocks and hunger vulnerability.

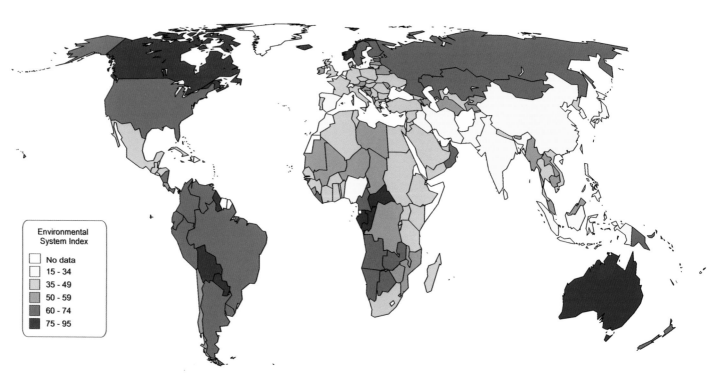

Environmental System Index
No data
15 - 34
35 - 49
50 - 59
60 - 74
75 - 95

Map 14.1. Environmental systems health, 2005.

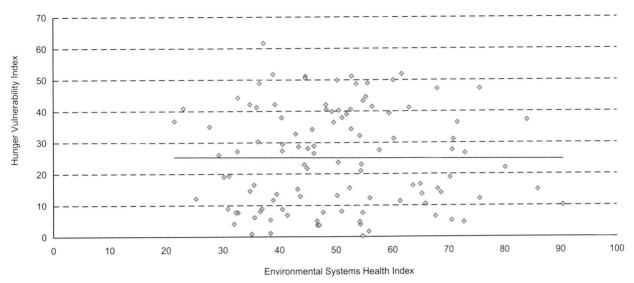

Figure 14.1. Hunger vulnerability and health of environmental systems. *Sources:* WHO 2008c; FAO 2008b; World Bank, PovcalNet, n.d.; Esty et al. 2005.

Figure 14.1 plots each country in terms of its score on the Hunger Vulnerability Index and its score on the environmental systems index. If more environmental resources automatically implied less hunger, we would expect the two indicators to be arranged in a downward sloping pattern. In fact, there is no clear pattern in figure 14.1. Countries with any level of environmental status can be found with almost any level of hunger vulnerability. To the extent that this measure of environmental status is accurate, the data show that resource stocks do not drive hunger outcomes independent of many other factors. The roles of politics in determining the distribution of resources and technology in determining their productivity may imply more about hunger than the level of resources itself. In low-income areas where environmental systems appear threatened (South Asia and East Africa on this map), the politics of distribution and of technology use may be especially critical.

# 15: Human Resources: Literacy

Human resources are as critical as natural resources in ensuring food security and nutrition. A population's productive capacity is affected by its education and health. Mapping literacy rates gives some indication of the state of human resources in a country. A later entry (chapter 31) will discuss health. The pattern of literacy rates in map 15.1 is similar to the distribution of hunger vulnerability in map 8.1. Figure 15.1 charts the relationship between hunger and adult literacy. While there are some countries, like Kenya, that have both high rates of hunger vulnerability and high rates of literacy, all countries with low literacy rates have uniformly high hunger vulnerability. Figure 15.1 shows that the Hunger Vulnerability Index is over 30 in almost all countries

where fewer than half of the adult population can read. Comparing figures 14.1 and 15.1 shows that literacy is more systematically related to hunger than the environmental systems measured in map 14.1.

Literacy and better nutrition may go hand in hand, because better education enables people to be more food-secure by improving their income earning potential and increasing their awareness of good nutrition. But factors that contribute to literacy also contribute directly to reduced poverty and hunger. As the arrows in figure 1.2 indicate, human assets influence national production, poverty and hunger even as those factors affect human resources, and all of them are influenced by political and economic power relations.

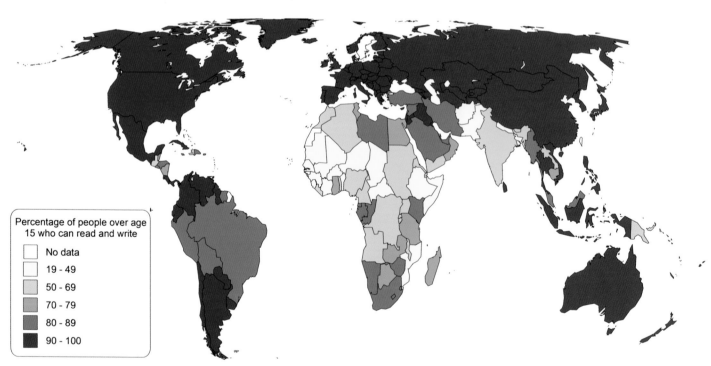

Percentage of people over age 15 who can read and write

- No data
- 19 - 49
- 50 - 69
- 70 - 79
- 80 - 89
- 90 - 100

Map 15.1. Adult literacy, 1992–2005.

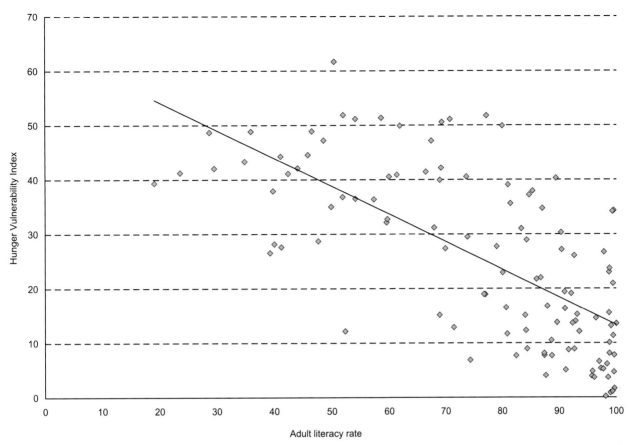

Figure 15.1. Hunger vulnerability and literacy. *Sources:* WHO 2008c; FAO 2008b; World Bank, PovcalNet, n.d.; World Bank, *World Development Indicators Online,* n.d.

# 16: Built Resources: Roads

An educated and healthy population with the capacity to convert natural resources into built resources can help a society be more productive. One measure of a country's built resources is its physical infrastructure. Map 16.1 shows the distribution of built resources around the world in terms of the road density of a country, measured as kilometers of roads per square kilometer of land area. Where the road network is small and in poor condition, many people may live in isolation from jobs, markets, and public services like education and health care. Such isolation is likely to contribute to poverty and hunger. Be that as it may, map 16.1 and figure 16.1 show little direct relationship between hunger and built resources.

There are many reasons why a link between roads and hunger may not be apparent. First, the quality of roads is not considered in these data. This tends to exaggerate the level of the road network in countries like India, where less than half of the roads are paved, compared to European countries like Germany where 99% of the road network is paved. Failure to account for quality means that some countries with high hunger vulnerability appear to have more built resources than they actually do. Another problem with this measure is that it does not account for the way in which a country's population distribution affects its need for roads. Large countries like Australia and Canada have vast unpopulated areas, often deserts, where roads would not contribute significantly to people's welfare. Low average road density in these countries obscures the fact that roads are in relatively good supply where people need them.

Despite these weaknesses, the map does reveal some relationship between built resources and hunger vulnerability. The link is clearer when we consider the overlapping effects of population density and roads.

Roads (km) divided by
land area (sq km)

- No data
- 0 - 0.24
- 0.25 - 0.4
- 0.5 - 0.9
- 1 - 1.4
- 1.5 - 8

Map 16.1. Road density, 2000–2005.

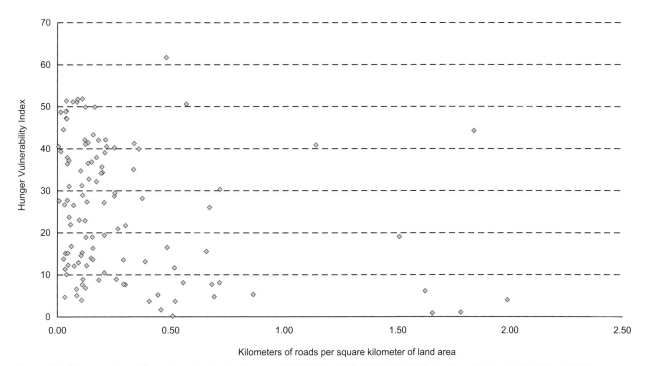

Figure 16.1. Hunger vulnerability and road network. *Sources:* WHO 2008c; FAO 2008b; World Bank, PovcalNet, n.d.; World Bank 2008a.

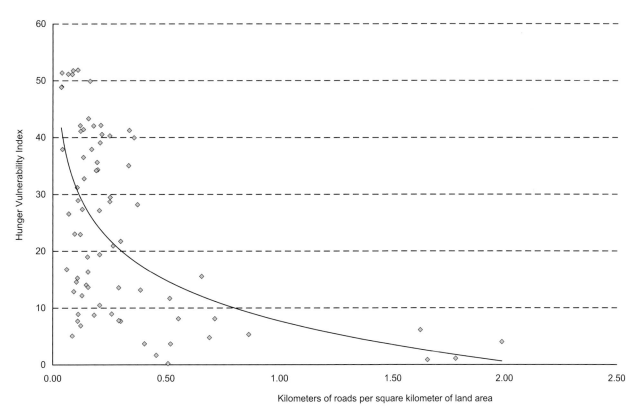

Figure 16.2. Hunger vulnerability and road network in countries with 25 to 250 people per square kilometer. *Sources:* WHO 2008c; FAO 2008b; World Bank, PovcalNet, n.d.; World Bank, *World Development Indicators Online,* n.d.

Densely populated countries that also have very limited road networks compared to their land area tend to have high hunger vulnerabilities (map 8.1). Many examples of such places can be seen in Eastern and Central Africa and in South and Southeast Asia. Figures 16.1 and 16.2 show a more clear relationship between hunger and built resources in countries where population density suggests a need for a tight road network than in countries in general. Together these data suggest that many countries with high rates of hunger vulnerability suffer from limited built infrastructure.

# 17: Change in Resource Base

The preceding maps indicate how specific resources like arable land, education, and roads are distributed around the world. The environmental systems health map attempted to capture key natural resources, but not human resources. In many ways, these kinds of assets can substitute for one another. A highly educated population, for example, may be very productive and prosperous despite scant natural resources. Many small countries like Singapore ensure food availability, despite having very little land, by using their substantial industrial and service sectors to generate income. To get an overall sense of the natural, human, and built assets available for meeting food availability goals, it is useful to combine various resources into a single indicator. We call this indicator the *resource base*.

One way of measuring the resource base is to use accounting methods to gauge how different types of assets are being built through investment or destroyed through extraction. In this approach, money spent on education could be seen as representing growth in human resources, while money earned from selling minerals or petroleum reflects the value of lost natural resources. Similarly, the value of investments in roads contributes to growth in human-made assets, while deteriorating roads reflect a decline in these resources. If gains in human and built assets were to exceed losses in natural resources, one could say that the resource base was growing. This growth should make it easier to achieve sufficient food availability.

The World Bank attempts to measure changes in a country's resource base through what it calls "adjusted savings." This measure sums up the value of expenditures in education and public and private investments and subtracts depletion of forests, minerals, and energy resources as well as deterioration of human-made assets. This gives a net value of the change in all these resources that is compared with a country's gross national income. Where the ratio of "adjusted savings" to income is negative, the country is "eating-up" its resource base, making it less and less able to produce

in the future. The more negative the ratio is, the less sustainable is the current level of production and consumption. Where the ratio is positive, the indicator suggests that a country's capacity to produce is growing.

This measure has the merit of combining different types of assets to indicate the broad resource base. However, the indicator is far from complete. It does not include important human assets like health or important natural assets like farmland. Because of these shortcomings, one must also look at indicators for specific resources such as health expenditures (see map 31.1) and arable land (maps 13.1–13.2) to get a more complete picture of how the resource base is changing.

In each major region of the world there is a mix of countries with growing, stable, and declining resource bases (map 17.1). Because this measure tries to capture change in a country's resources rather than the level of those resources, it suggests something about a country's capacity to support current consumption into the future, but it does not indicate the level of current consumption. Therefore, a rich country like Saudi Arabia that is depleting its resources is less likely to experience widespread hunger than a poor country like Tanzania, which may be nurturing growth in its resources but whose capacity to achieve food security remains weak. As a result, we find little relationship between the current distribution of hunger and the data displayed in map 17.1. Nonetheless, the map does yield some insights about future scenarios for world hunger.

The most rapidly declining resource bases are in oil-exporting countries that are extracting their energy resources more quickly than they are expanding human or built resources. In 2006, Kazakhstan, Iran, and Saudi Arabia all experienced double-digit rates of decline in their "adjusted savings." The income level in these countries implies that this loss in resources is not immediately threatening, but it does raise concerns about long-term sustainability.

It is more troubling to see declining resource bases in countries that currently have low incomes such as

Map 17.1. Change in resource base, 2006.

Change in resource base as % of national income

| | | |
|---|---|---|
| | No data | |
| | -70 to -11 | Rapid decline |
| | -10 to -1 | Moderate decline |
| | 0 to 4 | Stable |
| | 5 to 9 | Moderate growth |
| | 10 to 36 | Rapid growth |

Figure 17.1. "Bush fires? Never again!" Sign outside of Pa, Burkina Faso, urging citizens to preserve the country's natural resources by not setting fires. These signs were first erected during the Thomas Sankara government (1983–87) as part of the "Three Struggles" campaign against bush fires, deforestation, and wandering livestock. The signs were also part of the regime's efforts to combat illiteracy through promotion of national languages. This sign is in the Jula language.

Angola (−38%), Bolivia (−24%), and Zambia (−15 ). Maps 31.1 and 13.1–13.2 show that these countries are also experiencing low investments in health and declining land availability. Countries like these that are currently failing to meet food availability requirements and are also depleting their resource base will find it increasingly difficult to eradicate hunger. In contrast, countries such as India, Kenya, and Namibia are currently experiencing hunger but appear to have growing resource bases and may therefore be increasingly able to meet their food availability targets in the future. Of course meeting food availability goals, as we have seen, does not ensure that people eat well.

# 18: Climate Change

The surface temperature of the earth increased by 0.6° C (1°F) during the last century. Most of this warming is attributed to the burning of fossil fuels (oil, gas, coal), which has produced higher concentrations of carbon dioxide ($CO_2$) and other heat-trapping gases in the atmosphere (IPCC 2007). Climate change could have potentially enormous impacts on global agricultural production and food availability (map 18.1), as well as on people's livelihoods, their exposure to natural disasters, and their ability to access food. On the one hand, increased levels of $CO_2$ can stimulate plant growth and lead to higher levels of production. Yet, rising temperatures can also stress plants and stifle production. Since rainfall and temperature vary widely around the world, questions are being raised about the regional effects of climate change on food production and on dramatic weather events like floods, droughts, and hurricanes. Will some regions be more affected than others? How vulnerable are the populations of dif-

ferent regions to climate change? What is the capacity of rural and urban populations to cope with the severe weather events associated with climate change? Will they have the means to invest in new technologies and pay potentially higher prices for food?

To answer some of these questions, researchers have developed climate impact models that investigate the relationship between temperature changes, socioeconomic conditions, and agricultural production (Fischer et al. 2005; Hitz and Smith 2004; Parry et al. 2004). The results of this research were analyzed and presented in a report produced by the Intergovernmental Panel on Climate Change (IPCC 2007). Regarding the relationship between climate change and food insecurity, the report presents a number of findings that point to heightened vulnerability, especially in Sub-Saharan Africa. First, the report suggests that the impact of climate change on food production will vary regionally. Higher $CO_2$ levels and temperatures will likely result

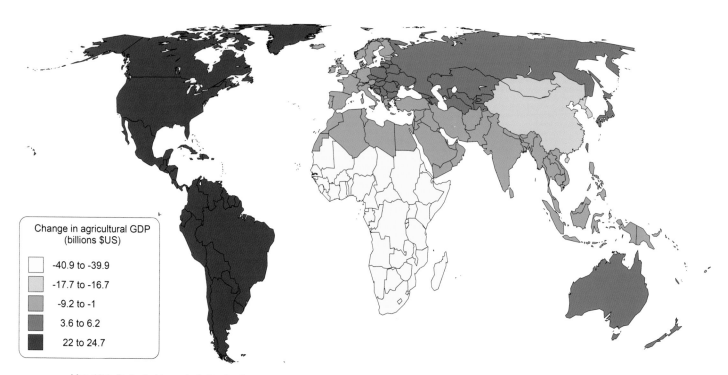

Map 18.1. Projected impact of climate change on agricultural production, 2000–2080, CSIRO Model.

in higher grain yields in higher latitudes but reduced yields in lower latitudes. However, regional differences in technical options and in projected rainfall complicate these climate and agricultural change simulations. Such differences help to explain why the Americas do not show this pattern (map 18.1) (Easterling et al. 2007, 280). Second, the impact of frequent and extreme climate events brought on by climate change portend to be more important for hunger than changes in average temperature and rainfall levels. That is, the weather-related events that can push poor and vulnerable households into hunger are likely to increase. Third, the modeling studies suggest that although climate change will increase the number of hungry people, socioeconomic development paths have a greater impact on hunger vulnerability. The IPCC report estimates that the number of hungry people will increase by between 40 and 170 million by 2080. The figure varies depending on the socioeconomic assumptions used in the model. In general, rapid economic growth and reduced regional differences in per capita incomes lead to lower numbers of people at risk of hunger. Conversely, persistent inequalities in regional per capita incomes and slow economic growth generate greater undernourishment. In the latter case, a regional shift in hunger will occur from Asia to Sub-Saharan Africa.

The findings of the IPCC panel advance our understanding of the complex relationships between global warming, climate change, and societal impacts. But the models and simulations are vague about the broader social and political-economic contexts that make people vulnerable to environmental shocks in the first place. For example, the socioeconomic variables like per capita income are simply imposed in the model, so the analysis reveals little about the underlying and ultimate causes of hunger vulnerability. As a result, there remains much uncertainty about how climate change might interact with social and political processes to affect access to food and food production.

There are many other gaps in knowledge that cloud our understanding of anticipated social impacts and vulnerabilities. For example, the effects of $CO_2$ and climate change have focused on cereals but not on other crops typically cultivated by the poor in developing countries. More research is needed on the effects of $CO_2$ and climate change on pests, weeds, and disease and how these interact with food availability and human health. The scale of research also needs to shift to the regional level to determine the effects of climate change and variability on crop production. In addition, the Intergovernmental Panel on Climate Change notes that the assumptions regarding trade and economic growth found in the integrated assessment models "were poorly tested against observed data." Finally, the impact of climate change on smallholder production is difficult to assess because of the complexity of farming systems, characterized by crop diversity, intercropping, and agropastoralism, which sharply contrast with the monocultural farming systems that drive the climate impact models (Boko et al., 2007; Easterling et al. 2007).

Perhaps the greatest uncertainty is how societies will adjust over time to climate change. Rapid rates of urbanization in Sub-Saharan Africa and Asia suggest that hunger will become increasingly an urban phenomenon that will require social policies aimed at increasing employment and entitlement programs that will reach at-risk populations. The current focus in global environmental change research on human vulnerability to climate change (Boko et al. 2007) should help policy makers design programs that will reduce the susceptibility of people to food shortages in specific political and economic contexts.

# Technology

A country's ability to meet its food availability needs hinges, in part, on the technology it can apply to its resource base. Technology is the manner in which resources are used to achieve some goal. Improved agricultural technology contributes to food availability by increasing agricultural production. This can alleviate hunger by reducing food prices to consumers and increasing the incomes of farmers. Other kinds of technology can create off-farm employment that can improve people's ability to buy food. A society could combine human resources that have little application in agriculture (knowledge of computer programming) with technologies that are not targeted to agriculture and generate income to buy an abundance of food. Thus, the total breadth and depth of a country's technological resources influence its capacity to meet food needs. Agricultural technologies have a particular role in hunger problems in places where many people rely on farming for their livelihoods. The maps 19.1–19.4 represent an effort to reveal the distribution of technological capacity around the world.

Due to data limitations, our measures of technology focus on technical knowledge and processes for discovery that come from developed countries. We have no measures to map the local technical knowledge that farmers and small-scale manufacturers in developing countries may be generating and applying to good effect. The data that are available indicate how well countries may be able to participate in the revolutionary expansions of information technology and biological research that are now occurring.

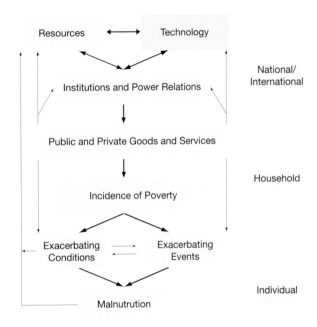

# 19: College and University Enrollment and Research and Development

The greater the share of a population with higher education, the greater the country's ability to use, adapt, and invent appropriate technologies. Map 19.1 shows college and university enrollment as a share of the population that is locally considered to be college-aged. In the wealthy regions of the world, opportunities for college education are widely available and enrollment rates are high, usually over 60%. In middle-income regions of the world, like Latin America, enrollment rates are somewhat lower (20%–50%). Meanwhile, in low-income regions like East Africa, college enrollment rates are under 5%. In these countries public spending on higher education is low, households usually lack the funds to pay for private colleges, and most people can never go to college. Indeed, low rates of primary school enrollment and of secondary school completion make university education unlikely in much of Africa. Most countries with such low rates of college enrollment will be unlikely to participate fully in the development and application of modern technology.

Like the countries of Africa, India and China also have very low rates of college enrollment. However because of their huge populations, the absolute number of college-educated people in these countries is quite high. Moreover, because of the larger scale of higher education, these countries have developed centers of excellence in teaching and research on specific topics. Thus, they are more likely to participate in and contribute to technology development than other countries in the global South. For example, Chinese agricultural researchers have developed genetically modified cottonseeds that compete with those produced by the US agribusiness giant Monsanto. China's technological and commercial achievement makes it a world leader in the high-tech arena of biotechnology.

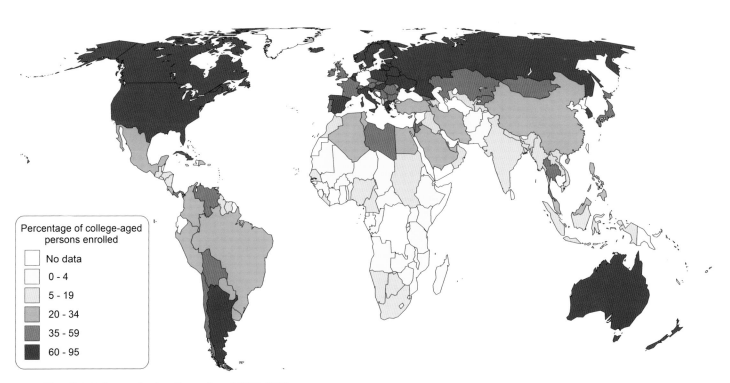

Percentage of college-aged persons enrolled

- No data
- 0 - 4
- 5 - 19
- 20 - 34
- 35 - 59
- 60 - 95

Map 19.1. College and university enrollment, 2000–2006.

## LOCAL RESEARCH AND DEVELOPMENT

In order to contribute to the creation or adaptation of technology to solve local problems, college graduates need to be working locally. Unfortunately, many college graduates from developing countries are drawn away from their homes to work in developed countries. Map 19.2 shows the number of practicing technicians and scientists actively engaged in research to develop new technologies per million inhabitants in the country. We would expect countries with greater representation of such professionals to be more able to use technology to convert their national resources into income that can ensure food adequacy.

Map 19.2 looks very similar to map 19.1. Countries with very low college enrollment rates necessarily have a very low share of their populations involved in formal research and development. There is likely to be considerable research occurring in these countries that is not reflected in the map. Many farmers in Africa and Southeast Asia may select seeds for specific traits and test crop varieties on their own farms. But without larger-scale research initiatives, their findings are unlikely to be spread to improve conditions widely. The limited presence of formal researchers in these countries suggests that little is being invested in technology development

and that the pace of innovation and diffusion of locally appropriate technology will be slow. Again, with their large populations, India and China may achieve more success in innovation because the small share of the population involved in technology amounts to a large number of people in those countries.

## SHARE OF NATIONAL INCOME INVESTED IN RESEARCH AND DEVELOPMENT

In order to be effective, researchers need tools and inputs that can be prohibitively expensive. Outfitting a laboratory typically costs millions of dollars. Countries that invest only small amounts in technology development are unlikely to have many technical innovations. Map 19.3 reports the annual investment that is devoted to research and development around the world. The higher this investment level is, the more funds available to researchers and the more productive we would expect them to be. The variable is calculated as a share of national income. For countries that have low incomes, the absolute value of these investments will be low even if the investment rate reported is quite high. Again, the pattern of the map is similar to that for college enrollments and for practicing researchers. Where there are few colleges and few researchers there

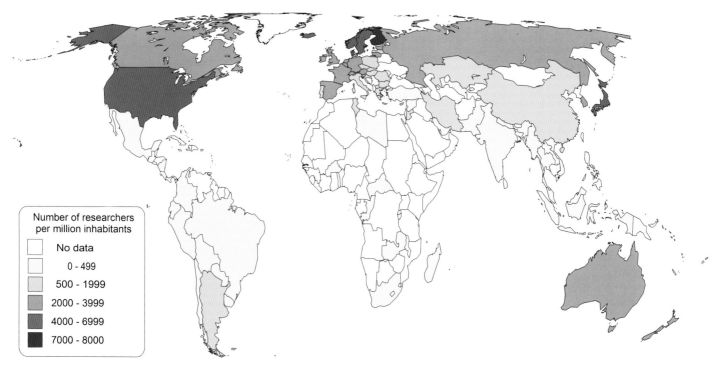

Map 19.2. Research in the world, 1996–2004.

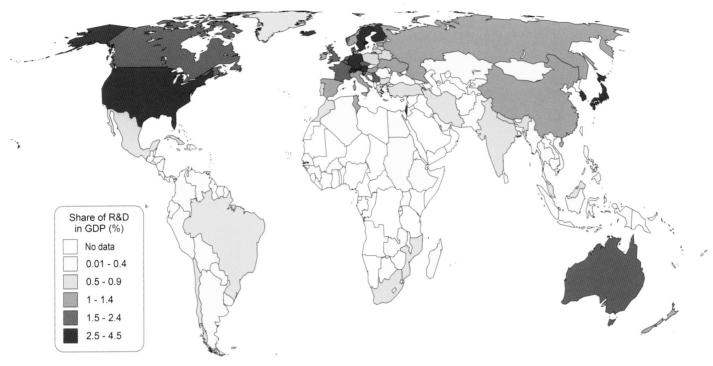

Map 19.3. Investment in research and development, 2000–2006.

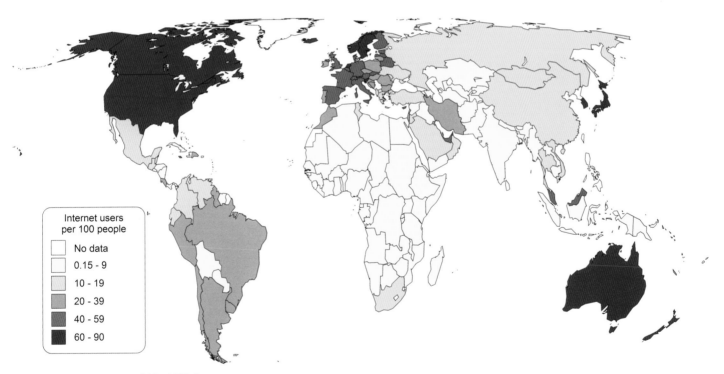

Map 19.4. The digital divide, 2003–6.

are few institutions through which the government or private agencies can invest in research. The reverse is also true; where there is little public or private investment in technology development, trained people will not be able to engage in research.

## THE DIGITAL DIVIDE

The Internet is a key tool for researchers of all kinds and is a powerful means of spreading new knowledge and skills to a population. The greater the Internet accessibility in a country, the easier it is to develop new technology and the more rapidly that technology can be spread to benefit the population. Map 19.4 captures the "digital divide" that separates countries with broad access to information technology from those without it. Once again, the split between the global North and the global South is clear.

These four maps are remarkable for their uniformity. Paucity in one measure of technology is a good predictor of poverty in the others. These indicators of technological capacity are mutually reinforcing: High rates of college enrollment imply that a large supply of researchers is being trained. Higher spending on R&D implies that locally trained researchers will be able to remain in their country of origin and that foreign scientists will be attracted to that country. A relatively large core of trained and employed researchers can stimulate Internet use and provide locally relevant content.

While countries that lack these forms of technology may be rich in indigenous knowledge that is not measured here, such countries are still likely to be excluded from the economic opportunities that revolutions in information and communications technology have created in many regions.

Not surprisingly, these measures of technology also mimic the distribution of income globally. Higher-income regions of the world have the capacity to invest in education and in R&D. Consequently, it is these regions that have the greatest capability to harness their resource base to generate income and control food. The unfortunate implication of these maps is that those regions that experience the most hunger and therefore have the greatest need to convert their resources into food appear to be the most limited by their technology base.

# 20: Agricultural Technology: Fertilizer

A technology is useful when it can sustainably enable people to produce more. When rural producers are empowered with better farm technologies they are able to grow more on their land and enjoy less vulnerability to hunger. Improved farm technologies can increase the availability of food, potentially lowering its price for urban consumers while raising the incomes of producers. Unfortunately, it can be as difficult to put sustainable technologies in the hands of low-income farmers as it is to develop them.

The use of chemical fertilizers is not a new technology, but as map 20.1 shows, fertilizer use around the world is very uneven. In most African countries less than 10 kilograms of fertilizer are applied per hectare of arable land, while the use rate in the United States is over 100 kilograms and in China it is almost 400 kilograms per hectare. Table 20.1 gives some evidence of the impact of these low levels of fertilizer use in Africa. As a region, Sub-Saharan Africa consumes far less fertilizer per hectare than other regions and also experiences yields for cereal crops and returns to labor in farming that are much lower than those enjoyed in other regions. Indeed, cereal yields in Africa are less

**Table 20.1. Fertilizer use by region**

| Region | Fertilizer use (Kg/HA) | Cereals yield (Kg/HA) | Return to labor (US $/year) |
|---|---|---|---|
| Developed market economies | 119 | 49,038 | 23,081 |
| Asia and Pacific | 174 | 28,049 | 423 |
| Latin America and Caribbean | 92 | 26,666 | 2966 |
| Middle East and North Africa | 79 | 24,478 | 2140 |
| Sub-Saharan Africa | 13 | 13,357 | 327 |
| World | 101 | 32,389 | 695 |

Note: Fertilizer data are for 2002. Cereals yield data are a 2003–5 average. Labor returns are for 2003. Fertilizer use excludes manures. Return to workers is in constant (2000) US dollars.

Source: FAO 2006b.

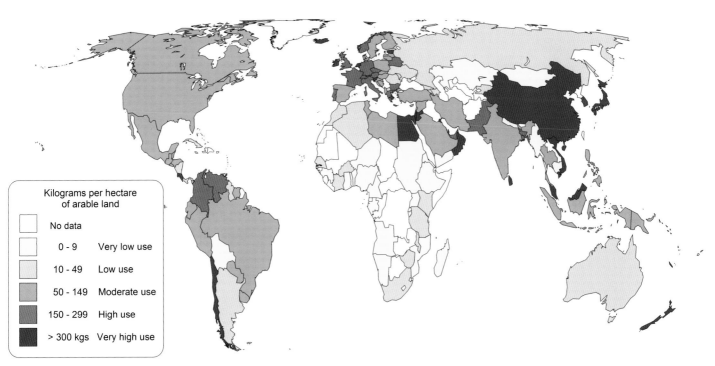

Kilograms per hectare of arable land

| | |
|---|---|
| No data | |
| 0 - 9 | Very low use |
| 10 - 49 | Low use |
| 50 - 149 | Moderate use |
| 150 - 299 | High use |
| > 300 kgs | Very high use |

Map 20.1. Fertilizer consumption. Includes nitrogenous, potash, and phosphate fertilizers, but excludes manures, 2004–5.

**Box 20.1. New Rice for Africa (NERICA)**

A breakthrough in rice development took place in the mid-1990s at the Africa Rice Center in Cotonou, Benin, one of 15 centers in the Consultative Group on International Agricultural Research funded by aid donors and private foundations. Scientists there successfully crossed African rice (*Oryza glaberrima*) and Asian rice (*Oryza sativa*) species to produce hybrid varieties called NERICA that are both hardy and high-yielding. The seeds are popular because yields are higher than local varieties even without fertilizer application. When fertilizers are applied, output doubles. The new varieties also mature two months earlier than local varieties, which shortens the difficult "hungry season" when food from the previous harvest is finished and the new harvest has yet to take place. The wonder seeds are drought and pest tolerant, grow well in difficult soils, and compete vigorously with weeds. There are currently 18 named varieties of NERICA cultivated by African farmers on some 150,000 hectares—a tiny proportion of the potential area. According to the Africa Rice Center, the main bottleneck slowing the diffusion of this new technology is seed supply (www.warda.org). There are too few government agencies and private firms producing sufficient amounts of seed for farmers to grow. But there are additional problems that hinder farmers' access to the NERICA varieties. Despite their great interest in the new seeds, farmers lack funds to buy the seeds and fertilizers to participate in this potential green revolution. Credit is scarce partly because both government and aid donor investments in rural development have plummeted over the past two decades (Digger 2007; Naylor and Falcon 2008). Poor farmers also lack the political clout to demand agricultural credit, improved roads, and a steady supply of seeds and fertilizers to rural areas. Despite promising breakthroughs in agricultural technologies and the enthusiasm of African farmers to adopt them, the new seeds are out of their reach and hunger vulnerability persists.

Figure 20.1. Farmers obtaining fertilizers on credit from a village cooperative in northern Côte d'Ivoire to grow cotton and food crops. Credit is deducted from the farmer's earnings when he sells cotton to the cooperative after harvest.

than half those in almost any other region. Differences in natural fertility are part of the reason for low yields and incomes in African agriculture, but use of fertilizer could compensate for many natural deficiencies.

Potentially useful technologies may go unused for many reasons. In the case of fertilizer in Africa, some farmers are unconvinced of the benefits. Many others, however, would like to use fertilizers but lack access to the technology. Remote farmers can find it difficult and costly to get to markets where fertilizers can be purchased. Frequently, they lack the cash to buy fertilizer when it is needed and have no options to borrow money or buy on credit with repayment after their harvest. While technologies may exist that can help people reduce their hunger vulnerability, access to technology usually depends on good information about its benefits, a good distribution system to make it accessible, and access to credit to make it affordable. Many of these supporting factors remain absent in the places where hunger vulnerability is greatest (box 20.1).

# Institutions and Power Relations

Institutions and power relations occupy a pivotal place in our model. On the one hand they shape how resources and technology are developed and utilized. On the other hand, they influence the distribution of goods and services. Subsidies that US farmers receive for growing cotton, sugar, and corn are one example of how national institutions (USDA commodity programs) and domestic power relations (politically influential farm organizations) affect the development and utilization of resources and technology. Land seizures in Zimbabwe and food price controls in Venezuela are two examples of domestic power relations working through national institutions to restructure production and distribution within countries. Power relations also operate between countries and shape the options that governments have in dealing with hunger issues. Internationally, institutions like the World Bank and International Monetary Fund alter the distribution of public goods and services through the terms they impose on developing countries seeking financial assistance. These conditions can entail unpopular cuts in spending for social and economic programs and can unleash economic shocks that governments would not allow in the absence of external pressure. Thus inter-national power relations and institutions have a significant impact on how national governments alleviate or exacerbate hunger vulnerability.

Maps 21.1–25.2 explore how differences in power between countries can affect hunger and then consider the role of domestic power relations on hunger vulnerability.

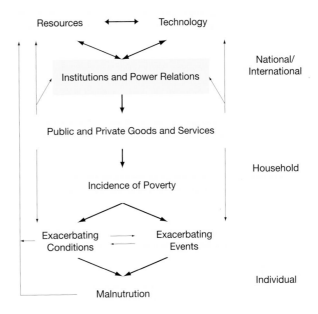

# 21: Colonialism and Neocolonialism

Not only are some countries more susceptible to hunger than others, but some are also more limited in their policy options for tackling it. Differences in the power of governments to independently chart their country's hunger and poverty strategies are rooted in their colonial history. European colonialism deeply affected the societies and economies of many of the countries that today experience hunger vulnerability. Colonialism is both a political economic system and an ideology that allows a country to assert control over a people and territory beyond its boundaries. Historically colonial power was exerted for the purported benefit of the subject population, but colonialism benefited the colonizing power to the detriment of the colonized.

European colonies covered 10% of the world's land area in 1750. By 1914 this territory expanded more than three-fold to encompass 35% of the world's land area (map 21.1). Colonial possessions brought prestige to even small countries like Belgium and Portugal, whose African colonies significantly enlarged their land area. King Leopold's personal colony, deceptively named the Congo Free State, was 76 times the size of Belgium (Hochschild 1999). By adding Angola and Mozambique to its colonial possessions, Portugal could proudly proclaim to the world that "Portugal is not a small country" (figure 21.1).

Colonial authorities sought to control the natural resources and labor of their colonies and to reorient markets to serve the interests of European industries, consumers, and settlers. In Southeast Asia, the Netherlands instituted the "culture system," which required farmers to plant one-fifth of their fields in cash crops like coffee and sugar that were of interest to Dutch companies (Bagchi 1983, 71–73). In Eastern and Southern Africa, the British forced indigenous peoples to move onto reserves so that European settlers could occupy the most productive lands. These reserves, known as Tribal Trust Lands, Communal Lands, and Native

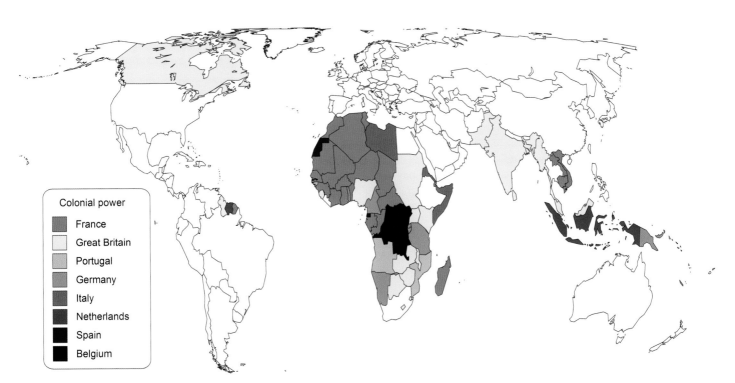

Map 21.1. European colonies, 1914.

Figure 21.1. Portuguese government propaganda map showing the size of its colonies in comparison to the land area of Portugal. The title reads: "Portugal is not a small country." Enrique Galvão, *Lithografia de Portugal* (Lisbon, ca. 1935), private collection.

Reserves, were notoriously overcrowded and characterized by poor soils and insufficient rainfall. Poverty, hunger, and taxation forced men and women to migrate to settler plantations and mines to work as wage laborers.

In French West Africa, the colonial state coerced rural populations to produce cash crops and work for a pittance in settler logging concessions and coffee and cocoa plantations. French policies in Côte d'Ivoire required savanna farmers to plant a specified area in cotton and then sell it for low prices in markets dominated by French merchants (figure 21.2). The little money earned from these transactions was used to pay taxes imposed by the colonial state on adults. This "head tax" had to be paid in French francs. The colonial state also required communities to fill a certain number of sacks with maize, rice, and millet. If a village did not have sufficient quantities of these crops to deliver, it had to buy them in a neighboring village. In this way, colonialism impoverished households and made them vulnerable to poor harvests associated with drought or pest invasions.

Some of the taxes collected in colonies were reinvested in those regions, but colonial investments shaped colonies according to European priorities rather than local interests. For example, governments supported research to develop cash crops of interest to European merchants and industries. Food crops important to the diets of local populations received little attention. Similarly, colonial investments in transportation infrastructure were designed to move goods from the interior to the coast, where they could be easily shipped to Europe, not to promote regional economic development.

The regime of forced labor, forced cultivation, dispossession, and taxation intensified during the first and second world wars and the Great Depression of the 1930s. In Africa, famine became common in areas where, prior to colonialism, it had been rare (Watts 1983). Sen (1983) chronicles the devastating famines that occurred in British India, where the entitlements that prevented famine in the past were eroded under colonial rule.

The legacy of authoritarian rule and the orientation of production and markets to serve the colonial powers is apparent in the continuing dependency of many former colonies on the export of primary products (map 33.1) and the large numbers of people living in

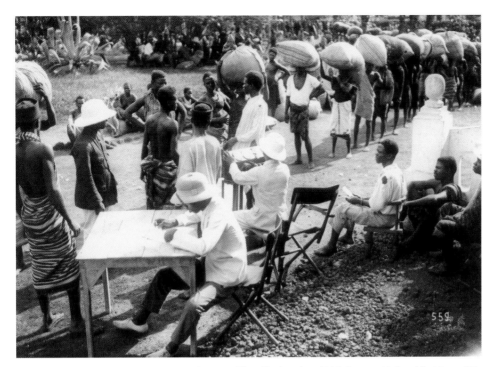

Figure 21.2. Forced cotton deliveries in Tiébissou, Côte d'Ivoire, circa 1915. *Source:* National Archives of France, Overseas Section.

extreme poverty (map 27.1). Former colonies tend to trade with their former colonial rulers, producing and exporting low-value primary goods, which makes them particularly vulnerable to deteriorating terms of trade for their exports (maps 34.1 and 35.1). Strong political ties also bind former colonies and colonial powers, as is evident in bilateral aid flows of France and Great Britain (table 21.1). The top 10 recipients of both British and French aid in 2004–5 were overwhelmingly former colonies.

Having inherited economies that were designed to serve a foreign country rather than local peoples, many former colonies have fragile economic systems that have been vulnerable to instabilities and crises. Often, economic crises have driven these countries to seek the aid of international financial institutions (IFIs) that have provided loans only on the condition that borrowers adopt a set of specific policies aimed at reducing government spending, increasing exports, and "freeing up" markets. In many countries these policies may have deepened poverty and increased hunger vulnerability

**Table 21.1. Top recipients of French and British official development assistance, 2005**

| Rank | French aid | British aid |
|---|---|---|
| 1 | Nigeria | Nigeria* |
| 2 | Congo Republic* | Iraq* |
| 3 | Senegal* | India* |
| 4 | Morocco* | Bangladesh* |
| 5 | Iraq | Zambia* |
| 6 | Madagascar* | Afghanistan |
| 7 | Algeria* | Tanzania* |
| 8 | Cameroon* | Ghana* |
| 9 | Mayotte* | Congo, Democratic Republic |
| 10 | Tunisia* | Sudan* |

*Former colony

Source: OECD 2007

(Stiglitz 2003). The efforts of IFIs like the World Bank and International Monetary Fund to dictate national social and economic policy resemble the power of European governments during the colonial era. This resemblance has led many critics of the World Bank and IMF to label their interventions as "neocolonial."

# 22: Debt and International Power Relations

As the map of colonization shows, many regions of the world were officially under the control of foreign states during the first half of the twentieth century. Today, these former colonies often remain in weak positions of power compared to high-income countries and former colonial rulers. These power relations are frequently displayed in the economic interactions between lenders and borrowers. Financial relations among countries are relevant to a discussion of hunger, because they can limit a government's choices for addressing hunger and poverty.

The countries of Africa and Asia that gained their independence after the Second World War often found themselves needing massive investments in health, education, and infrastructure in order to enable their people to move out of poverty. Domestic resources, including export earnings, were used to fund these expenses, but over time governments increasingly sought foreign lenders to provide finance for their investments. Loans from private multinational banks, foreign governments, and international financial institutions were to be used for public investments that would generate income to improve local livelihoods and repay the foreign debt. However, by the 1980s many former colonies found themselves with more debt than they could hope to repay.

The debt crisis emerged for a number of reasons. In some cases borrowed funds were used for consumption rather than productive investment. Often poor oversight and corruption put in private hands the money that had been borrowed for public investment. As a result, when repayments had to be made, there was no additional income with which to pay. In other cases loans had been made when a borrowing country's exports were selling at high prices, but when those prices dropped, the country could no longer make payments. This problem was compounded by rising costs of imports like oil. In still other cases, countries had invested the borrowed money in projects that they expected would yield high returns, but were in fact misguided and unsustainable. The causes of the debt crisis were many, but the result was that debtor countries often found themselves seeking assistance from two international financial institutions: The World Bank and the International Monetary Fund (IMF).

The World Bank and IMF were established by a United Nations mandate in 1944 to provide finance to member governments. The World Bank was designed to lend governments funds for long-term investments like construction of highways and ports. The IMF was to provide short-term loans to soften the effect of transitory financial crises. These institutions are governed by the member nations, with each country holding voting rights in proportion to its financial contribution.

In the 1980s, the World Bank and the IMF began making loans to governments so that they could pay off their debts to other creditors. This was appealing to the borrowers in part because the new loans would have lower interest rates than the original ones and because they provided more time over which to repay. However, because the two lenders believed that the policies of the borrowers were a cause of the debt crisis, they made their loans on the condition that the borrowing governments reform their economic policies. Borrowers found that their debts and the need to find funds with which to repay them resulted in foreign powers exerting influence over their domestic policies. The so-called "structural adjustment" lending typically required countries to devalue their currency, deregulate their domestic markets, liberalize their foreign trade, and reduce their public spending.

Map 22.1 shows the size of a country's debt compared to its annual export earnings. The ability to repay loans depends on exports because debtor countries usually have to repay in US dollars or euros. In general, if a country's total debt is worth less than 100% of its annual export earnings, lenders consider it to be in a good position to repay eventually. Once the value of

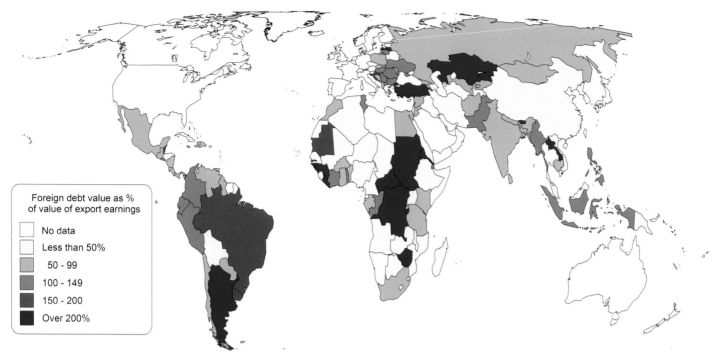

Map 22.1. Foreign debt burden, 2006.

debt grows to above 150% of annual exports it is un-likely that the country will ever be able to pay it off without assistance.

Another indicator of a country's debt burden is the share of its export earnings that must be paid each year to service the debt. Annual debt service payments of over 30% of export earnings are considered unsustainable. By the 1990s, many countries had debts that were unsustainable, and a growing movement was calling for debt forgiveness. As figure 22.1 shows, about 25% of the 110 indebted countries for which we have data had debt service payments of over 30% of their export revenues in 1990. Only about half of the indebted countries had debt service payments under 20% of their export earnings that year.

Recognizing that much of the debt held in poor countries could never be paid, the World Bank and IMF began the Highly Indebted Poor Countries (HIPC) Initiative. This program aimed to forgive some of the debt of borrower countries in order to bring the remaining debt down to a sustainable level. As the figure shows, the share of countries with debt service payments of over 30% of export earnings has fallen significantly since the 1990s. Many of those countries that continue to have high debt burdens are not clas-

sified as "poor" and have been ineligible for the debt forgiveness program. Debt forgiveness was conditional on governments adopting specific policy reforms and committing to use the funds that were freed from debt repayment for social services. The relatively low values of the total debt which map 22.1 shows for Mozambique and Tanzania reflect debt reductions through the

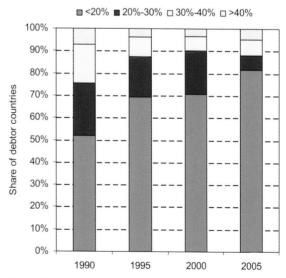

Figure 22.1. Debt service as a share of export revenue for 110 debtor countries. *Source:* World Bank, *World Development Indicators Online*, n.d.

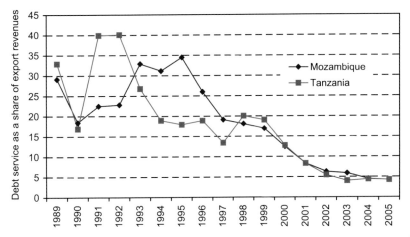

Figure 22.2. Declining debt service under HIPC initiative. *Source:* World Bank, PovcalNet, n.d.

HIPC initiative. Figure 22.2 reveals the declining debt service in these two countries under the HIPC initiative. The reductions in debt that these countries have received required their governments to develop domestic policies in concert with international organizations. Whether or not those policies prove to be beneficial, the process used to establish them reflects the limits on the power of governments in poor countries compared to wealthy ones.

# 23: Political Freedoms

Just as power relations among countries limit the options some governments have for addressing hunger, so power relations within countries can determine which strategy among the available options a government chooses. One might expect governments that are more accountable to their people to be more active in combating hunger. If so, there should be a relationship between hunger and the degree of democratic freedom in a country. Are there lower rates of hunger vulnerability in countries where people are free to join political organizations, run for office, and elect individuals who represent their interests? Do people enjoy greater food security in countries where they can express themselves freely in newspapers and in public demonstrations?

Some theorists like Amartya Sen argue that a relationship does exist between famine prevention and the level of political rights and civil liberties in a country (Sen 1990). In democratic societies, elected officials would be ousted from office if a famine occurred under their watch. The case of India, the world's largest democracy, is illustrative. Mass starvation has been deterred in that country since political independence in 1947 in part because political leaders would be publicly disgraced if such an event occurred. An independent press has also played an important role in politicizing famine conditions. In short, there exists in India an "anti-famine political contract" between elected governments and vulnerable but politically active popula-

Figure 23.1. Political campaign for regional council elections in Korhogo, Côte d'Ivoire, July 2002.

tions. This contract is enforced by popular protests, a vigilant press, and the self-interest of politicians and civil servants to avoid explosive political issues like mass death from starvation (de Waal 1997). While democratically elected leaders may be especially inclined to avoid famine, it is less clear that political rights imply greater initiative or effectiveness in preventing chronic hunger.

Freedom House, a research institute funded by public and private sources, developed two measures of political freedom (political rights and civil liberties) that can be used to explore whether a relationship exists between hunger vulnerability and political freedoms. Freedom House defines these two measures in the following way:

> Political rights enable people to participate freely in the political process, including the right to vote freely for distinct alternatives in legitimate elections, compete for public office, join political parties and organizations, and elect representatives who have a decisive impact on public policies and are accountable to the electorate. Civil liberties allow for the freedoms of expression and belief, associational and organizational

rights, rule of law, and personal autonomy without interference from the state. (Freedom House 2006)

Freedom House conducts annual surveys and uses the results to give countries a rating between 1 and 7 to reflect their level of political rights and civil liberties. Countries receiving a score of 1 have the highest levels of political rights and civil liberties, while those with a score of 7 have the lowest amount. Ratings are given for each measure and then averaged to obtain a country's overall freedom rating. A country's freedom rating is then placed into one of three categories: free (rating 1–2.5), partly free (rating 3–5), and not free (rating 5.5–7). The map of political freedoms is based on these ratings (map 23.1).

When we compare a country's freedom rating with its hunger vulnerability score, the relationship turns out to be very weak. Figure 23.2 shows very little clustering around the trend line. This finding suggests that the existence of democratic institutions and civil liberties as measured here is not sufficient to make political leaders willing and able to reverse poverty and hunger in their countries (de Waal 1997). Indeed, our maps and graphs suggest that leaders in a "free" country like In-

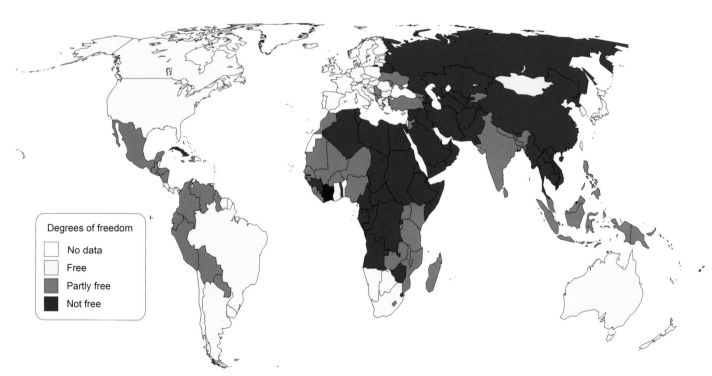

Degrees of freedom

- No data
- Free
- Partly free
- Not free

Map 23.1. Political freedoms, according to Freedom House 2008.

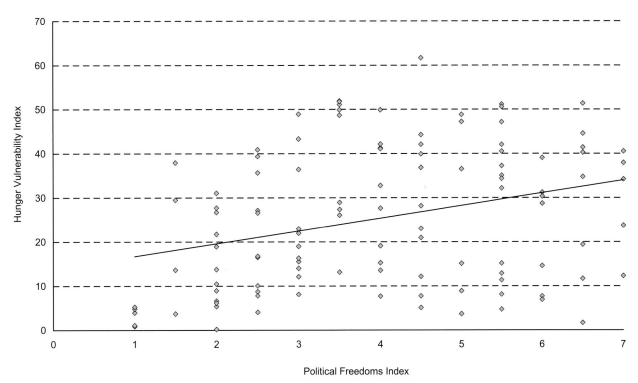

Figure 23.2. Political freedoms and hunger vulnerability. *Note:* Lower values imply more political freedom. *Sources:* WHO 2008c; FAO 2008b; World Bank, PovcalNet, n.d.; Freedom House 2006.

**Table 23.1. Selected freedom and hunger vulnerability ratings**

| Country | Freedom rating | Hunger vulnerability |
|---|---|---|
| Brazil | Free | Medium |
| India | Free | Extremely high |
| Mali | Free | Extremely high |
| United States | Free | Low |
| Colombia | Partly free | Medium |
| Ethiopia | Partly free | Extremely high |
| Tanzania | Partly free | Very high |
| Algeria | Not free | Medium |
| China | Not free | High |
| Côte d'Ivoire | Not free | High |
| Libya | Not free | Low |

Source: Freedom House 2006.

dia can find it politically tolerable to have millions of people experience extremely high levels of hunger vulnerability (compare map 8.1 with map 23.1). Table 23.1 provides examples of the random relationship between hunger vulnerability and relative freedoms.

The fact that a large share of a country's people can be left in extreme deprivation despite their civil rights and political liberties may reflect how difficult it is to eliminate chronic hunger and poverty. Careful planning and decisive action by leaders may be sufficient to mobilize domestic and international resources to prevent famine. Poverty and hunger are more complex problems. Even where there is the political will to end chronic hunger, governments in poorer countries may not have the capability to finance costly food security programs. This is probably the case in Mali, where extreme hunger vulnerability and high political freedoms exist side by side. Mali has a per capita gross national income of $1100 (PPP) and ranks among the poorest 20 countries in the world. The existence of political freedoms is helpful in preventing famine from taking place in that country, but it is not sufficient to address the much deeper and wider problem of poverty.

# 24: Income Inequality

Differences in political power are often related to income inequality. Income inequality and poverty are related but distinct concepts. In this atlas, "poverty" refers to a condition of absolute deprivation. We have classified people as "in poverty" if they live on less than $2.00 per day (PPP). "Inequality," in contrast, refers to relative rather than absolute conditions. Measurement of income inequality usually tries to gauge the distribution of income, not its level.

One common way to measure inequality in a country is to calculate the share of national income held by the richest and poorest 10% of the population (maps 24.1 and 24.2). If incomes were distributed equally, the poorest 10% would control 10% of the nation's income and the richest 10% would hold the same share. Of course, in the case of complete equality, there would be no richest or poorest 10%, since everyone would have the same income. In general, the larger the share of national income held by the richest 10% or the smaller the

share of income held by the poorest 10%, the greater the income inequality in a country. (See box 24.1 for descriptions of other measures of inequality.)

The implications of income inequality for people's material well-being depend on the amount of income that is available to share. In terms of distribution, the countries of Scandinavia (Sweden, Norway, and Finland) are very similar to those of South Asia (Pakistan, India, Bangladesh). In both regions the poorest 10% claim a large share of national income compared to other countries, while the richest control a smaller share than elsewhere. However, since gross national incomes (GNI) in South Asia are about $2000 (PPP) per capita compared to $30,000 (PPP) per capita in Scandinavia, the poorest 10% in India, Pakistan, and Bangladesh are at vastly greater risk of hunger than are their counterparts in Sweden, Norway, and Finland. Similarly, the poorest 10% in Ethiopia control 4% of the country's income, while the poorest 10% in the United States hold

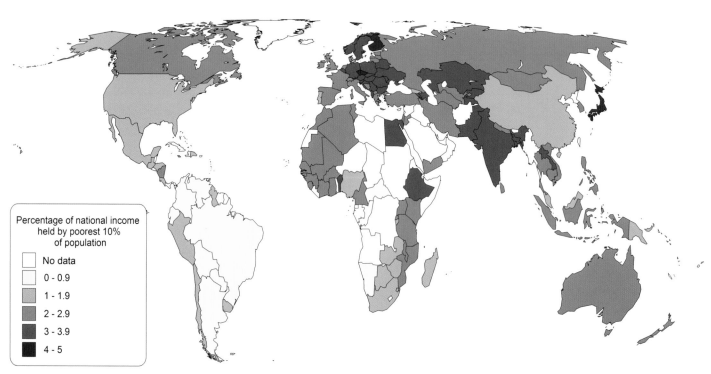

Percentage of national income held by poorest 10% of population

☐ No data
☐ 0 - 0.9
☐ 1 - 1.9
☐ 2 - 2.9
☐ 3 - 3.9
☐ 4 - 5

Map 24.1. Income share held by poorest 10%, 1992–2005.

## Box 24.1. Measuring Inequality: Lorenz Curves and Gini Coefficients

The maps on inequality reveal the share of national income controlled by the poorest 10% of a country's households and the share held by the richest 10%. A more general presentation of income distribution can be plotted using a Lorenz curve. The Lorenz curve diagram (figure 24.1) shows the percentage of national income on the vertical axis and the percentage of households or population ordered from lowest to highest income on the horizontal axis. The Lorenz curve maps the cumulative share of income held by progressively larger shares of the population. A Lorenz curve will always end in the upper right corner of the diagram, with 100% of households controlling 100% of national income and will always begin in the lower left corner. The shape of the curve suggests something about distribution.

If a Lorenz curve followed a 45-degree line from the lower left to upper right, the country would have perfectly equal income distribution. The poorest 10% would control 10% of national income; the poorest 40% would control 40% of national income; and so on. The curved solid line in figure 24.1 shows a more common distribution. Here the poorest 10% have about 5% of national income; the poorest 40% have about 20% of national income; and the richest 20% have 30% of national income. The greater the inequality in distribution, the larger the area between the 45-degree line and the actual Lorenz curve, labeled A in this figure.

The gini coefficient, or gini index, is a measure of inequality based on the Lorenz curve. It is simply the area A in figure 24.1 divided by the area (A + B). The gini coefficient must fall between zero and 1, with zero representing perfect equality and 1 representing a single individual holding all the national income. Sweden, a country with fairly even income distribution, has a gini index of 0.25. In South Africa, where a history of racial injustice has created highly uneven distribution, the gini coefficient is 0.58. The United States has a gini value of 0.41, while Ethiopia's is 0.30.

The gini coefficient is a useful and widely applied summary measure of inequality. However, it does not indicate the status of people at the extremes of the distribution. Two Lorenz curves could imply the same gini coefficient but different shares of income to the poorest groups. The dotted Lorenz curve in figure 24.1 has about the same gini coefficient as the solid Lorenz curve, but the poorest households are shown to have a smaller share of national income with the dotted curve. A focus on hunger implies particular interest in people at the bottom of the income distribution, and concerns about economic and political power suggest emphasis on those at the top. With that in mind, the maps in this atlas show the share of income that goes to the poorest and richest 10% of households.

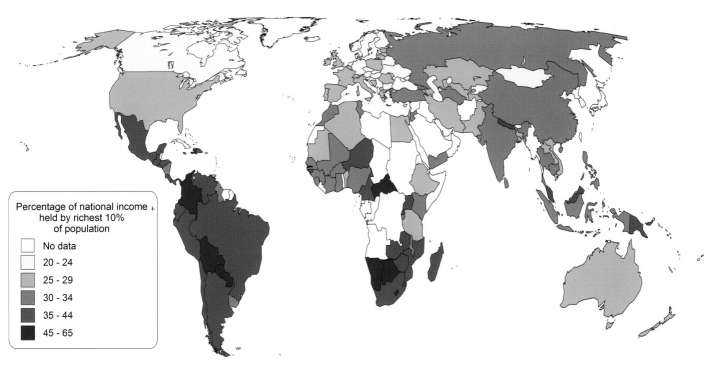

Percentage of national income held by richest 10% of population

No data
20 - 24
25 - 29
30 - 34
35 - 44
45 - 65

Map 24.2. Income share held by richest 10%, 1992–2005.

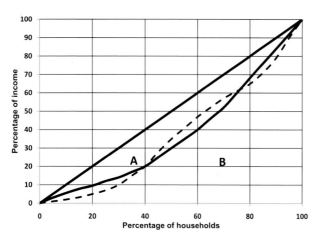

Figure 24.1. Lorenz curve diagram.

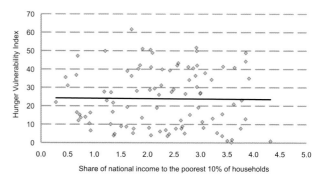

Figure 24.2. Hunger vulnerability and income inequality in the world. *Sources:* WHO 2008c; FAO 2008b; World Bank, PovcalNet, n.d.; World Bank, *World Development Indicators Online,* n.d.

less than 2% of their country's income. Since average income in the United States is about $40,000 compared to less than $1000 (PPP) in Ethiopia, the poorest Americans have considerably higher incomes than the poorest Ethiopians. (See box 7.1 for details on comparing incomes using purchasing power parity [PPP] exchange rates.)

Inequality can influence hunger vulnerability in at least two ways. First, if incomes are not very high, then unequal distribution of income will mean that the poorest people are too poor to get enough to eat. A second and less direct way in which inequality affects hunger is through the political system. If the rich control a large share of national income and the poor control a very small share, as in most of Latin America, it is unlikely that the poor will have much influence in political institutions. Inequalities in income distribution can translate into inequalities in political access. One would not expect governments that are more influenced by the rich than the poor to aggressively address the causes of hunger or to emphasize food and nutrition programs, since undernourishment is not a malady of the wealthy.

At first glance, the maps of income distribution bear little relationship to the mapping of hunger vulnerability. One can find countries like Ethiopia with high hunger vulnerability and low income inequality and countries like Chile with low hunger vulnerability and

dramatic inequality. This outcome can be explained by the influence of income level in addition to income distribution on hunger outcomes.

Figure 24.2 plots the data on hunger vulnerability and the share of income to the poorest 10% of the population in all countries for which information is available. The distribution in the diagram reveals no clear pattern. This is because in affluent countries high degrees of inequality can exist with low rates of absolute poverty. However, if we look only at those countries in which per capita incomes range between $5000 and $10,000 (PPP) per year, then a more distinct pattern appears. For these countries, the larger the share of national income going to the poorest, the lower the hunger vulnerability tends to be (figure 24.3). This relationship does not hold in the poorest countries, as the flat distribution of points in figure 24.4 shows. The level of income is so low in these countries that marginal increases in the share going to the poor (say from 2% to 3%) cannot influence hunger substantially. In these countries, alleviating poverty requires both increases in total income and more equitable distribution of it. Poverty alleviation can take place through government programs that provide jobs, health care, education, and food stamps to the poor. This channeling of public goods and services to assist the poorest requires that institutions and power relations be focused on reducing poverty and food insecurity.

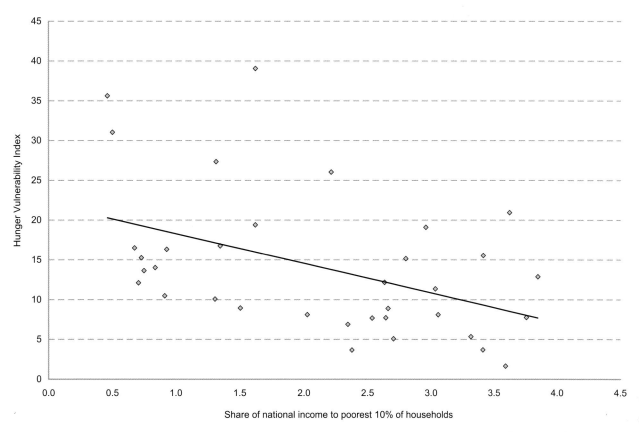

Figure 24.3. Hunger vulnerability and income inequality in countries with annual per capita incomes between $5,000 and $10,000 (PPP). *Sources:* WHO 2008c; FAO 2008b; World Bank, PovcalNet, n.d.; World Bank, *World Development Indicators Online,* n.d.

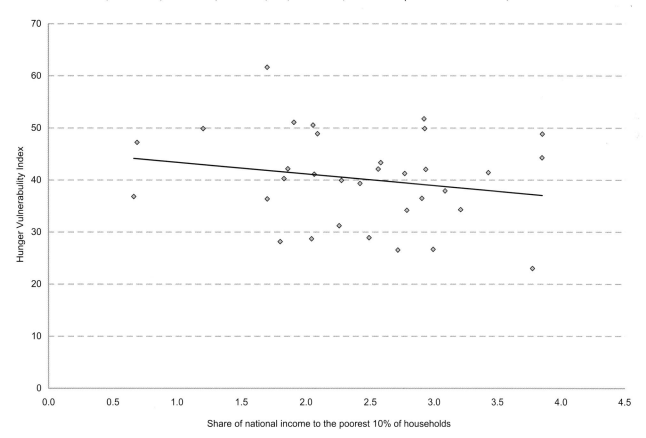

Figure 24.4. Hunger vulnerability and income inequality in countries with annual per capita incomes between $750 and $2,500 (PPP). *Sources:* WHO 2008c; FAO 2008b; World Bank, PovcalNet, n.d.; World Bank, *World Development Indicators Online,* n.d.

# 25: Gender Inequality

Women around the world play important roles in household economies. Whether as farmers in Africa, where they cultivate and prepare food for the household, or as urban income earners, women make significant contributions to household food security. Women's important economic contributions to households are in addition to their social responsibilities in bearing and caring for children. The precise contribution of men and women to household food production and purchases, health care and child care, will vary among cultures and socioeconomic groups. Yet studies from around the world show that increases in women's income and empowerment correspond to improved child nutrition. Since women are key in ensuring good nutritional outcomes, it is likely that places in which women are excluded from economic, political, and social opportunities will be disadvantaged in addressing hunger.

The Gender-related Development Index (GDI) created by the United Nations Development Program (UNDP) is a useful measure that illustrates the extent of gender inequalities across the world. The GDI indicates how women compare to men on three measures of human development: life expectancy, literacy, and income per capita. The GDI ranges from zero to one, with one representing gender equality and lower scores signaling greater inequality. Although women generally enjoy a longer life expectancy then men, they consistently rank lower when it comes to income and education levels. The GDI data show that in *every* country of the world, women experience lower levels of income and education than men. This reality suggests that women's ability to contribute to household food security is widely constrained by social discrimination.

Map 25.1 presents the gap in life expectancy between men and women in the world. In the more

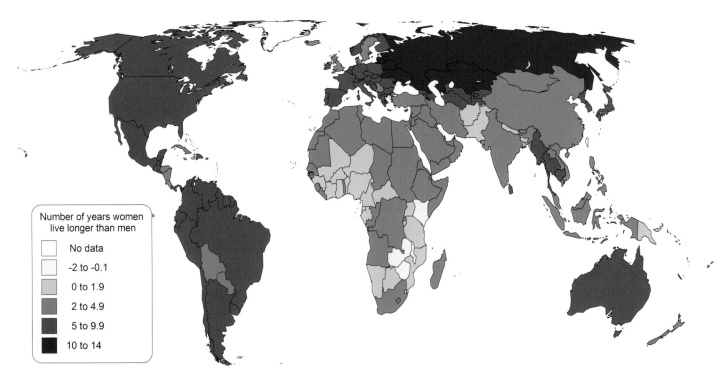

Number of years women
live longer than men

| | |
|---|---|
| | No data |
| | -2 to -0.1 |
| | 0 to 1.9 |
| | 2 to 4.9 |
| | 5 to 9.9 |
| | 10 to 14 |

Map 25.1. Gender differences in life expectancy, 2002–5 average.

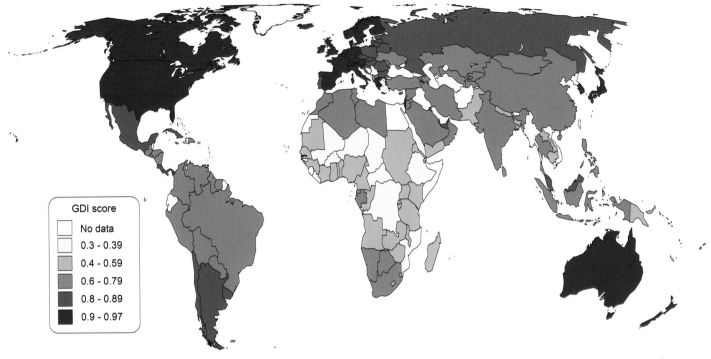

Map 25.2. Gender-related Development Index, 2005.

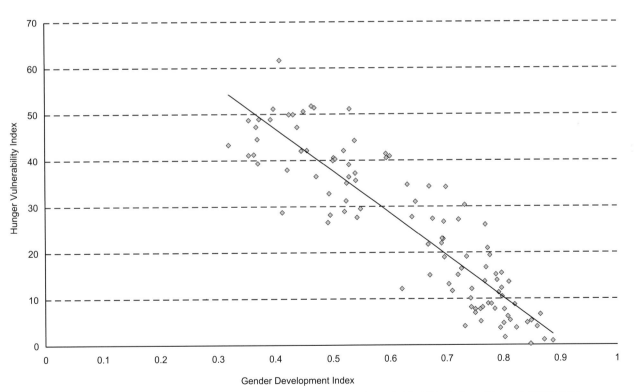

Figure 25.1. Gender inequality and hunger vulnerability. *Sources:* WHO 2008c; FAO 2008b; World Bank, PovcalNet, n.d.; UNDP 2008.

developed countries, women tend to live longer than men. This advantage is lost in the poorest countries of Sub-Saharan Africa, where women die much earlier. In some countries like Kenya and Malawi, disease (e.g., HIV/AIDS) greatly reduces women's life spans to below men's. These lost years for both men and women reduce household food security.

The geography of gender inequality (map 25.2) shows that disparities in human development among men and women are highest in Sub-Saharan Africa and South Asia. Strong resemblances exist between this map and the hunger vulnerability map (map 8.1) that suggest a relationship between women's status and hunger in the world. Indeed, when we plot gender inequality against the Hunger Vulnerability Index, a clear association becomes apparent (figure 25.1). The scatter diagram shows a very strong correlation between hunger vulnerability and gender inequality. The higher the gender inequality in a country, the higher is its vulnerability to hunger. Two clusters of countries stand out in figure 25.1. A food-insecure and low gender equality cluster appears on the upper left of the trend line. Sub-Saharan African countries make up 88% of this group. The second cluster is to the lower right of the trend line. The countries in this group rank better on the gender equity and Hunger Vulnerability Indexes. This cluster includes countries in Europe, North and South America, Southwest Asia, and Northern Africa. Within both clusters, increasing gender inequality tends to increase hunger vulnerability.

The South Asian countries of India and Bangladesh, with relatively low GDIs of 0.572 and 0.499 respectively,

Figure 25.2. Farmers eating a midday meal in their field in southern Burkina Faso.

also exhibit high hunger vulnerabilities. The UNDP believes that high gender inequality in India helps to explain the large numbers of undernourished people in that country (212 million in 2001–3) despite the overall declines in poverty rates in South Asia (UNDP 2005, 43).

The implications of gender inequalities of access for food security are striking in figure 25.1. The graph confirms the view that women's empowerment positively affects children's nutritional outcomes (Kennedy and Peters 1992).

# Poverty and Hunger

At the center of our hunger framework is household poverty. At the scale of the household we see how social, economic, and political processes interact to produce or reduce hunger vulnerability. In our conceptual framework, the prevalence and distribution of poverty are influenced by a country's resource and technology base and the political and market institutions that determine how resources and technology are used. These institutions operate at both the national and international levels. Poverty arises in political economies where access to public goods and services is insufficient, where too little is produced, and where the distribution of production is highly uneven across social groups. In the absence of entitlement programs that provide food safety nets, employment, and expanded production among smallholders, it is poverty that determines who is vulnerable to hunger. Examining the specific origins of poverty in a country or region requires detailed analyses of the politics, ideologies, and social relations that shape political economies, entitlements, and vulnerabilities. The hunger vulnerability framework situates poverty within this social and political constellation. The measurement of that poverty, its distribution globally, and its trends are explored in the next six maps.

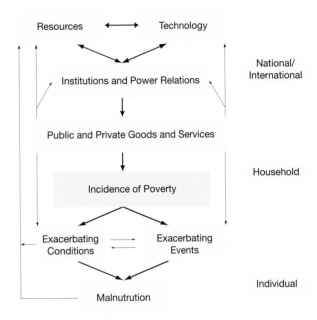

# 26: National Income per Capita

Are people in poor countries more vulnerable to hunger than people living in rich countries? Classification of countries as "rich" or "poor" is commonly based on measurement of gross national income (GNI) per capita. This is the measure used by the World Bank and other institutions to classify countries as high-income, upper-middle income, lower-middle income, or low-income.

GNI is calculated by assessing the market value of all goods and services produced by residents of a country. It includes both domestic production and income from abroad that is claimed by residents, but it excludes locally earned income that is sent to people living in a foreign country. Thus, the millions of dollars sent to Mexico each year by Mexican workers in the United States contributes to the GNI of Mexico (Porter 2006). Per capita GNI is simply total GNI divided by the population.

GNI per capita has a fairly straightforward definition, and because the information is recorded for almost all countries it is a convenient statistic to use. But GNI per capita can be a misleading indicator of people's prosperity for at least three reasons.

In the first place, there is the problem of income distribution. Whether a country's GNI leaves its citizens in destitution or in material comfort depends on how national income is distributed. Uneven distribution is a recurring theme in this atlas (see map 2.1 on food availability, maps 24.1–24.2 on inequality and hunger, and map 25.2 on gender inequality and hunger). Like the food balance sheets, which indicate how much food is available in a country but not who is consuming the available calories, GNI per capita tells us how much income is available in a country but not how that income is distributed across the population.

A second problem with GNI is that it values certain goods and services but not others. Because GNI relies on market prices to value things, it is weak at measuring gains and losses that don't appear in any market.

For example there is no market for public land, like national parks. If there is a toxic waste spill on such land, GNI will count the money that is spent on cleaning it up as income. Even if the incident does permanent damage to the environment, it can appear as an addition to GNI, since the environmental damage is outside the market. Similarly, there is no deduction from GNI when an activity reduces air quality or exhausts a natural resource. In a real sense, a country is poorer when its air quality deteriorates or its natural resources are destroyed, but these losses are often obscured in the GNI.

Just as costs that are outside of any market place are absent from GNI, so too is the value of nonmarket production. This production might be household tasks. If you clean your house, it is not a part of GNI. But if you hire somebody to do the exact same service, the activity is part of GNI. On a larger scale, if a rural person in Honduras produces $100 worth of vegetables which she eats and shares with her neighbors, it may not be caught in GNI since it does not enter the market. On the other hand, if she produces $100 worth of green beans that are frozen and exported, they will be counted as part of national income. The implication is that countries in which people tend to rely less on the market and more on informal exchanges will have a lower GNI than those that produce and consume at the same level but with more market exchanges.

A third problem in using GNI to measure relative wealth and poverty has to do with international price conversions. National income is initially measured in local currencies, like dollars in the United States, pesos in Mexico and rupiah in Indonesia. To make international comparisons, national income figures need to be converted into a common currency. The simplest way to do this would be to choose a currency to use as the common measure (say US dollars) and use the current exchange rates paid by banks to convert all values into dollars. One problem with this method is that exchange rates can vary from day to day, so the reported GNI could be misleading if the exchange rate on a particular day were not representative. For example, in February 2005 there were 9250 Indonesian rupiah to each US dollar, but by August 2005 the exchange rate was 10,750 rupiah per dollar, about 15% higher. The GNI in US dollars for Indonesia could differ by 15% depending on which exchange rate was used. To address the problem of exchange rate variability, GNI figures are usually converted using the average market exchange rates over an extended period. The "World Bank Atlas method" used by many international organizations is based on the average exchange rate for the prior three years.

Differences in the cost of living among countries present another problem for international comparisons of GNI. The costs of housing and of hiring labor tend to be much lower in developing countries than in developed ones. Because of these cost differences, a family may have a hard time making ends meet on US$12,000 a year in Switzerland or the United States, but might be comfortable on the same income in Swaziland or Bangladesh. To address these cost-of-living differences, values can be converted into a common currency using "purchasing power parity" (PPP) exchange rates (see box 7.1). These conversion factors are used to adjust

for cost differences so that any reported dollar amount would buy roughly the same amount of goods and services in any country. A person earning $12,000 (PPP) a year would support the same lifestyle whether she were in the United States, Swaziland, or Bangladesh. Thus PPP exchange rates, or "international dollars," lead to more meaningful international comparisons of GNI than methods using market rates. PPP conversions are used in map 26.1 and are also used in determining the share of population living on less than $2.00 per day (map 7.1).

One would expect higher income to imply lower hunger vulnerability. Despite all the problems with the use of GNI per capita as a measure of people's income, there is a close relationship between it and hunger vulnerability. One can see this by comparing map 26.1 with the hunger vulnerability map (map 8.1). Regions like South Asia and Sub-Saharan Africa and Central America all have low average GNI and high hunger vulnerability. This relationship is also shown in figure 26.1, which plots per capita GNI against the Hunger Vulnerability Index. The figure clearly shows that as average incomes fall, hunger vulnerability rises.

Just as figure 26.1 shows a general pattern of falling HVI with rising average incomes, it also reveals a great range of hunger vulnerability measures for any income

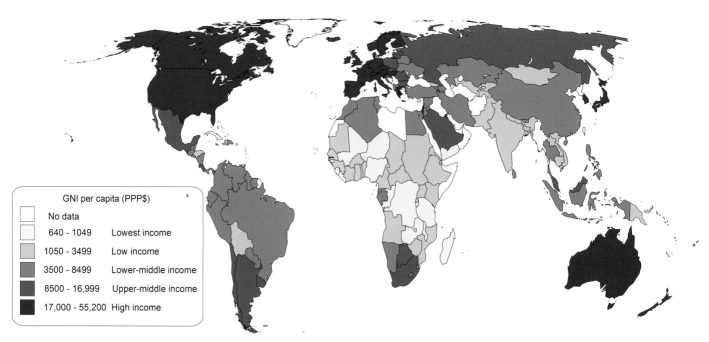

Map 26.1. Gross national income per capita, 2006.

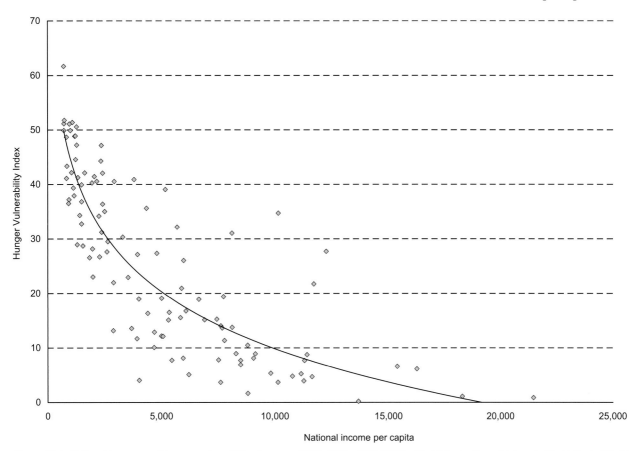

Figure 26.1. National income and hunger vulnerability. *Sources:* WHO 2008c; FAO 2008b; World Bank, PovcalNet, n.d.; World Bank, *World Development Indicators Online,* n.d.

level. Countries with incomes of $5000 to $7000 have HVI scores ranging from around 3 to over 30. For example, Uruguay and Botswana share similar GNI/capita levels, but the hunger vulnerability and poverty of their populations contrast sharply (table 26.1). When we use another measure of hunger such as child growth failure, countries in this income range have rates ranging from 7% to over 20% (figure 26.2).

The extent to which hunger vulnerability is related to average income depends largely on the relationship between average income and poverty rates. Figure 26.3 shows that as average income rises, the share of the population living on less than $2.00 a day falls. This relationship can also be seen by comparing maps 7.1 and 26.1. In countries with low average incomes, under $3500 PPP per year, poverty rates are inevitably high and hunger is a constant threat. In these countries, average incomes must rise to tackle poverty and hunger. As figures 26.1 and 26.3 show, many of the countries in the lower-middle-income category ($3500–$8500)

**Table 26.1. Similar income per capita, contrasting hunger vulnerabilities**

|  | GNI/cap ($PPP, 2006) | $2/Day poverty (%) | HVI |
|---|---|---|---|
| Botswana | 12,250 | 49 | 28 |
| Uruguay | 11,150 | 2 | 5 |
| Namibia | 8110 | 62 | 31 |
| Macedonia | 7610 | 3 | 4 |
| Kazakhstan | 7780 | 17 | 11 |
| Gabon | 5310 | 19 | 15 |

Source: World Bank 2008a; World Bank, PovcalNet, n.d.

achieve low rates of poverty and low hunger vulnerability, but many others fail to do so. Raising average income levels is clearly only part of the challenge of eradicating poverty and hunger.

When a country produces very little compared to the size of its population, poverty and hunger are inevitable. Thus, countries with very low gross national income per capita suffer high hunger vulnerability. While it may be necessary, raising average national income

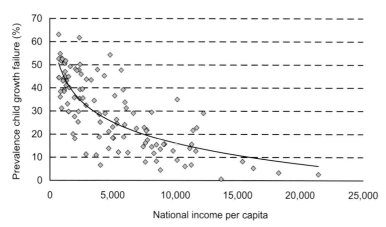

Figure 26.2. National income and undernutrition. *Sources:* WHO 2008c; World Bank, *World Development Indicators Online,* n.d.

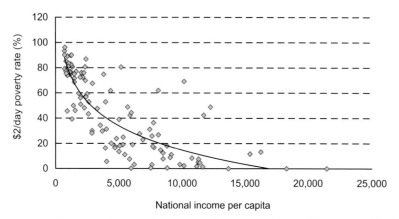

Figure 26.3. Poverty rate and national income. *Sources:* WHO 2008c; FAO 2008b; World Bank, PovcalNet, n.d.; World Bank, *World Development Indicators Online,* n.d.

is not sufficient for eradicating hunger. First, increases in national income must come through real increases in production, not simply accounting conventions that confuse resource depletion with production. More importantly, growth in average income can alleviate hunger only if the distribution of new income favors the poor. Income growth that enriches only those who are already in comfort cannot diminish hunger. Income growth that favors the poor is central to the eradication of poverty and undernutrition.

# 27: Extreme Poverty

People suffer hunger because they are poor and receive inadequate assistance. Regardless of national income, the more people who suffer poverty in a country, the greater the hunger problem will be. There is some relationship between national income and hunger vulnerability, because countries with lower incomes tend to have more people in poverty. However, we get a better picture of where people suffer poverty by looking at personal income rather than national income. For this reason, the United Nations, World Bank, and other international organizations collect extensive data on incomes and spending among people around the world. A key piece of information that emerges from these data is the number of people living on less than $1.25 per day. These people are considered to be in extreme poverty.

The $1.25/day poverty line is an international poverty line measured in purchasing power parity (PPP) dollars rather than US dollars. As explained in box 7.1, $1.25/day PPP refers to an amount of income that is equivalent in its purchasing power to having $1.25 in the United States in 2005. Individuals living in extreme poverty could consume no more than an American could buy with $1.25. No one can question that this describes extreme poverty.

In the Millennium Development Goals, the global community set a target of reducing the share of the population living in extreme poverty to half of its 1990 level by the year 2015. As illustrated in figures 7.2 and 7.3, extreme poverty rates have declined in some regions. Led by China, the rate of extreme poverty in East Asia fell from 55% of the population in 1990 to 17% in 2005 (table 27.1). This means that there were more than 557 million fewer people in extreme poverty in that region in 2005 than in 1990. However, progress in poverty reduction was uneven. In Sub-Saharan Africa there was little change in the share of the population in poverty. The number of people in extreme poverty actually rose by 92 million.

Maps 27.1 and 27.2 indicate progress that has been made toward the Millennium Development Goals for poverty, regions that are lagging, and places where the prevalence of extreme poverty is exceedingly high. Consistent with the table above, the maps show that rates of poverty are extremely high in South Asia and in Sub-Saharan Africa. India has made some progress in reducing the rate of extreme poverty. The poverty rate fell from 53% of the population in 1990 to 42% in 2005 (table 27.2). Overall, the rate of poverty in the South Asian region fell by 12 percentage points.

The greatest reductions in the $1.25/day poverty rate are found in China, declining from 60% in 1990 to 17% in 2005. Rates of extreme poverty in Pakistan fell from 65% to 23%, while Vietnam registered a decline from 64% to 21% and Indonesia's poverty rate fell from 54% to 22%. In most cases, countries that experienced reductions in poverty combined general growth in the economy (per capita GNI), with wide distribution of that growth.

Despite progress in reducing the rates of extreme poverty in the world, there were still 1.4 billion people living on less than a $1.25 a day (PPP) in 2005. Maps 27.3

**Table 27.1. Extreme poverty by region, 2005**

| World Bank Region | Share of population living on less than $1.25/day (PPP) (%) | | Number of people living on less than $1.25/day (PPP) (millions) | |
|---|---|---|---|---|
| | 1990 | 2005 | 1990 | 2005 |
| East Asia & Pacific | 55 | 17 | 873 | 316 |
| South Asia | 52 | 40 | 579 | 596 |
| Latin America & Caribbean | 10 | 8 | 43 | 46 |
| North Africa & Middle East | 4 | 4 | 10 | 11 |
| Europe & Central Asia | 2 | 4 | 9 | 17 |
| Sub-Saharan Africa | 58 | 51 | 299 | 391 |
| Low & middle-income countries | 42 | 25 | 1696 | 1377 |

Source: World Bank, PovcalNet, n.d.

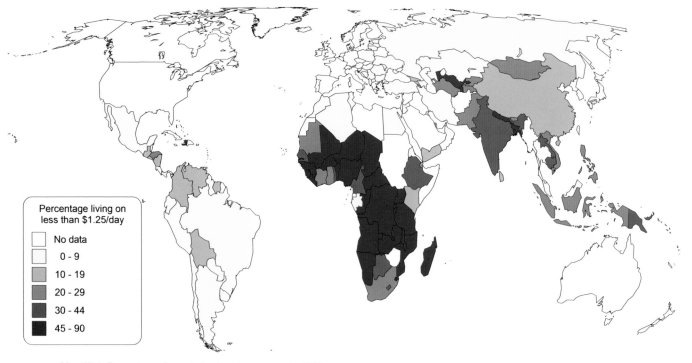

Map 27.1. Percentage of population in extreme poverty, 2005.

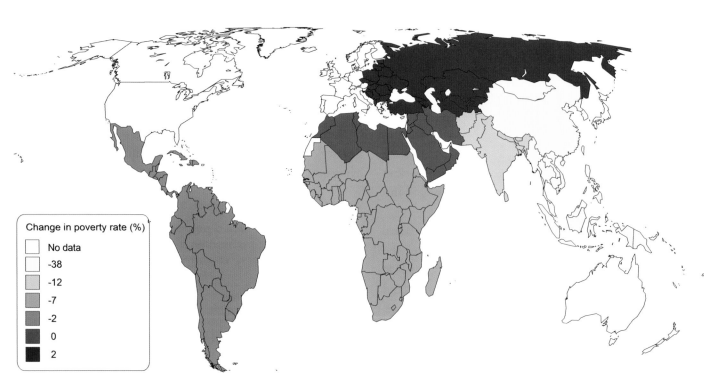

Map 27.2. Change in rate of extreme poverty, 1990–2005. *Note:* Change measured as the rate in 2005 minus the rate in 1990. Data are for regional aggregates.

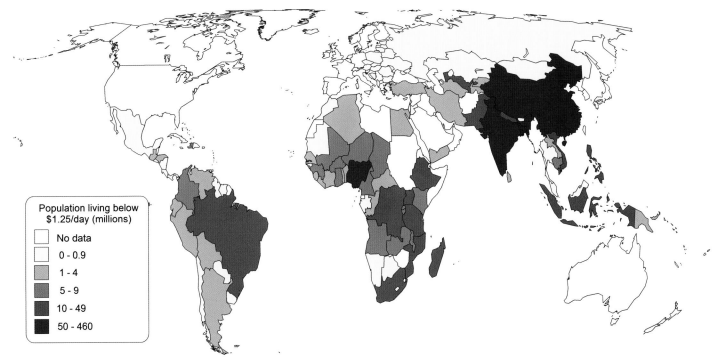

Map 27.3. Population in extreme poverty, 2005.

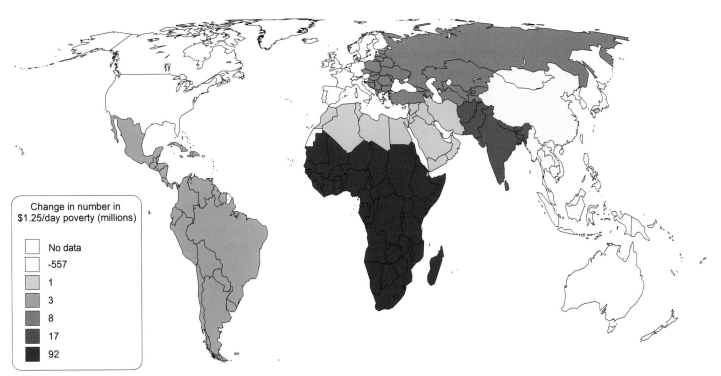

Map 27.4. Change in number of people in extreme poverty, 1990–2005. *Note:* Measured as the number of people in $1.25/day poverty in 2005 minus the number in poverty in 1990. Data are for regional aggregates.

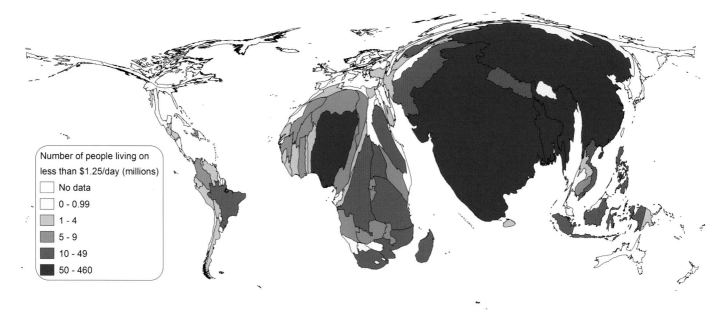

Map 27.5. Number of people living in $1.25/day poverty, 2005 (cartogram).

and 27.4 show where these people are found. Although poverty rates are falling in India and China, these countries still account for almost half of the people living in extreme poverty, as illustrated in the proportional area map (map 27.5). Between these two countries there are over 665 million people subsisting on less than $1.25 per day. Two additional countries, Nigeria and Bangladesh, account for another 167 million people in extreme poverty. These four countries hold about 60% of the world's extremely poor population. Economic growth that favors the poor in these countries could dramatically reduce the number of people in poverty and hunger globally.

Recent experience has been very uneven in China, India, Bangladesh, and Nigeria (table 27.2). While China has reduced both the rate of poverty and the number of people in poverty, India and Bangladesh have seen reductions in poverty rates with small increases in the numbers of people in poverty. Meanwhile both the

**Table 27.2. Share and number of people living in extreme poverty for selected countries**

|  | Share of population living on less than $1.25/day (PPP) (%) | | Number of people living on less than $1.25/day (PPP) (millions) | |
|---|---|---|---|---|
|  | 1990 | 2005 | 1990 | 2005 |
| China | 60 | 17 | 685 | 213 |
| India | 53 | 42 | 454 | 455 |
| Bangladesh | 67 | 50 | 75 | 76 |
| Nigeria | 49 | 64 | 46 | 91 |
| Average/total | 58 | 31 | 1261 | 835 |

Source: World Bank, PovcalNet, n.d.

share of the population and the number of people in poverty have risen dramatically in Nigeria. If India, Bangladesh, and Nigeria can stimulate economic growth that benefits the poor, then hundreds of millions of people could move out of extreme deprivation.

# Exacerbating Conditions and Events

Households in poverty are more at risk of hunger than others, but whether individuals in poor households suffer malnutrition depends on many conditioning factors. Unequal distribution within households, for example, could be an exacerbating condition that makes it more likely that someone in a poor household will go hungry. The presence of many people who cannot work because of their health or their age would be another exacerbating factor, as would the presence or absence of clean water. In addition to conditions like these that are somewhat stable, the experience of some dramatic event can also push people in poor households into hunger. Such events include catastrophic experiences like natural disasters or violent conflicts, but they can also be as mundane as drops in the price of an agricultural commodity.

The next maps explore some of these conditioning factors that can influence the effect of household poverty on an individual's nutrition. To give some order to the long list of factors that could place impoverished people at higher risk for hunger, maps 28.1–32.4 present conditions that appear in households (like sanitation), while conditions that define entire countries (like trade patterns), and disastrous events like floods or wars are presented in maps 33.1–38.2 and 39.1–40.1.

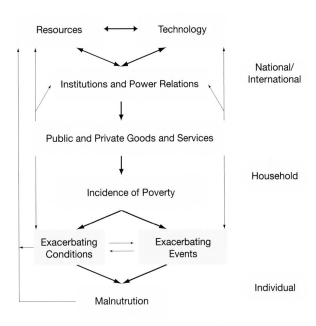

# 28: Dependency Ratio

Hunger often occurs in households in which there are few productive workers and many dependents. The dependency ratio indicates how many nonworkers each worker supports. It can be measured for a household or an entire country as in map 28.1. The ratio is calculated by dividing the percentage of a population that is unproductive by the percentage that is economically active. The unproductive population refers to people under the age of 15 and over 65. The productive labor force encompasses people between the ages of 15 and 65. This interpretation assumes that children are in school rather than working and that the elderly are retired from fulltime work.

The higher the dependency ratio, the more difficult it is for the workers in a household or country to provide for the population. If the dependency ratio is 1.0, every working-age person in the population must support herself and one other person. A dependency ratio of 0.5 means that there is one dependent for every two

workers. In Ethiopia 44% of the population in 2007 was under the age of 15, while 3% was 65 years or older (table 28.1). In this case, 47% of the population would be viewed as unproductive and 53% as economically active. Ethiopia's dependency ratio of 0.89 (47/53) represents a significant burden for a population with limited incomes and precarious harvests.

If families tend to have more children, the share of the population that is under 15 will be high, causing high dependency ratios. Large families are common among farming households, since children often contribute labor to farm tasks and may be the sole support of parents in their old age (figure 28.1). Parents' anticipation of mortality among children also contributes to high fertility as does poor access to family planning services for families that seek them. Thus, in developing countries that are mostly agricultural, 40% to 50% of the population is under 15 years old and dependency ratios are close to 1 (table 28.1). In developed countries,

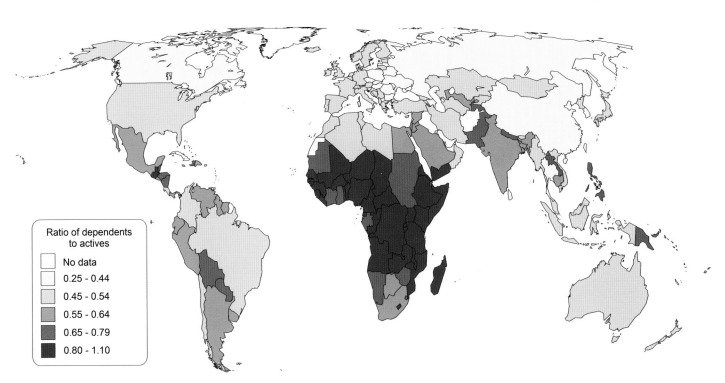

Ratio of dependents
to actives

No data
0.25 - 0.44
0.45 - 0.54
0.55 - 0.64
0.65 - 0.79
0.80 - 1.10

Map 28.1. Dependency ratio, 2007.

**Table 28.1. Dependency ratios and composition of dependents**

| Country | Dependency ratio | Share of population under 15 years (%) | Share of population over 65 years (%) |
|---|---|---|---|
| China | 0.41 | 21 | 8 |
| Ethiopia | 0.89 | 44 | 3 |
| India | 0.59 | 32 | 5 |
| Malawi | 1.00 | 47 | 3 |
| Mali | 1.08 | 48 | 4 |
| Mexico | 0.56 | 30 | 6 |
| Norway | 0.51 | 19 | 15 |
| Peru | 0.58 | 31 | 6 |
| South Africa | 0.57 | 32 | 4 |
| Sweden | 0.54 | 17 | 18 |
| United Kingdom | 0.51 | 18 | 16 |
| United States | 0.49 | 21 | 12 |
| Zambia | 0.96 | 46 | 3 |

Source: World Bank 2008a (2007 data).

dependency ratios are dominated by the elderly rather than the young. For example, Sweden and Peru have comparable dependency ratios, although in Sweden roughly equal numbers of people are under 15 and over 65, while in Peru almost a third of the population is under 15 and only 6% is over 65.

As map 28.1 indicates, dependency ratios are generally lower in richer countries than in poorer ones.

Developing countries that have aggressive family planning programs (China, for example) also have fairly low dependency ratios. Comparison of the map of dependency ratios with the map of hunger vulnerability reveals close correspondence but not necessarily causality. High dependency ratios in poor countries are a product of high fertility rates in rural households that are vulnerable to hunger for many reasons. As illustrated by map 12.1 (population growth), however, it is not at all clear that high fertility causes hunger vulnerability. Indeed, India, Bangladesh, and Myanmar represent countries that have achieved lower fertility and relatively low dependency ratios, but remain mired in widespread poverty and hunger vulnerability. High fertility, hunger, and poverty are often found in combination, but lowering fertility (and dependency ratios) does not necessarily lead to lower hunger or poverty levels.

While fertility may not cause hunger vulnerability, high dependency ratios do present challenges. In many countries, AIDS has significantly raised dependency ratios by reducing the population of working people. In Zambia, for example, there are 600,000 children orphaned by AIDS, meaning one child in eight is with-

Figure 28.1. Man with his two wives and nine children in Katiali, Côte d'Ivoire. Seven of the household members are under the age of 15 and none of the adults are over 65. With 58% of the household population classified as unproductive, and 42% active, the dependency ratio is high (1.38) for this rural household. In reality, the children do not go to school and contribute to the household farm economy by planting, weeding, and harvesting crops.

**Table 28.2. Children orphaned by AIDS**

| Country | Population under 15 years (millions) | Children orphaned by AIDS (millions) | AIDS orphans as a share of children under 15 (%) |
|---|---|---|---|
| Botswana | 0.72 | 0.10 | 13 |
| Ethiopia | 31.5 | 0.65 | 2 |
| Malawi | 4.5 | 0.56 | 12 |
| Nigeria | 59.84 | 1.20 | 2 |
| South Africa | 14.4 | 1.40 | 10 |
| Tanzania | 16.2 | 0.97 | 6 |
| Uganda | 12.5 | 1.20 | 10 |
| Zambia | 4.6 | 0.60 | 13 |

Source: World Bank 2008a; UNAIDS 2008.

out a birth parent (table 28.2). Nigeria, South Africa, Tanzania, and Uganda each have nearly a million or more AIDS orphans. Having lost their parents, these children are now dependent on others. The burden on the surviving adults is clearly worsening and not all will cope.

The following comments about a visit to Zambia in 2003 by Stephen Lewis, the UN secretary general's special envoy for HIV/AIDS in Africa, give a clear image of how AIDS is affecting the ability of households to provide for dependents.

We entered a home and encountered the following: to the immediate left of the door sat the 84 year old patriarch, entirely blind. Inside the hut sat his two wives, visibly frail, one 76, the other 78. Between them they had given birth to nine children; eight were now dead and the ninth, alas, was dying. On the floor of the hut, jammed together with barely room to move or breathe, were 32 orphaned grandchildren ranging in age from two to sixteen.[12]

This increasingly common condition of households composed primarily of "dependents" can only result in increasing hunger vulnerability. The members of the household described above, and countless like it, will almost surely face hunger.

# 29: HIV/AIDS

The loss of life from HIV/AIDS is staggering. More people have died from HIV/AIDS than in all the wars and disasters on earth over the past 50 years (FAO 2004a, 10). Some two million people died of HIV/AIDS in 2007 alone, with three-quarters of the deaths taking place in Sub-Saharan Africa. Across the world, 33 million people are infected with the virus; 22 million of these live in Sub-Saharan Africa (UNAIDS 2008). Map 29.1 shows that Southern Africa is the hardest hit region in the world. Infection rates exceed 20% of the adult population in Lesotho, Botswana, and Swaziland. They are also high in Eastern and Western Africa. Table 29.1 lists the 14 countries in the world with adult prevalence rates over 5%. All these countries are located in Sub-Saharan Africa.

HIV/AIDS directly affects household food security by reducing the number of agricultural workers and urban wage earners. People between the ages of 15 and 49, the most productive segment of the labor force, are the most frequent victims of the AIDS epidemic. When these people die, they leave behind young children and elderly parents who are much less productive. Meanwhile, healthy members of HIV/AIDS-afflicted families spend more time caring for sick people, making it harder to procure food and earn income. It is estimated that by 2020, HIV/AIDS will reduce the agricultural labor force by one-fifth in Namibia, Botswana, Zimbabwe, and Mozambique (UNAIDS 2006, 101). This reduced workforce threatens to increase hunger vulnerability in HIV/AIDS-afflicted households.

HIV/AIDS also drains households of limited financial resources on account of increased medical and funeral expenses. This means that there is little money left for investing in agriculture, education, and family health care. Some households are forced to sell crops, cattle, and even land to care for family members with HIV/AIDS. Distress sales, combined with declining investments in agricultural and human development, can significantly heighten food insecurity in both the short and long terms.

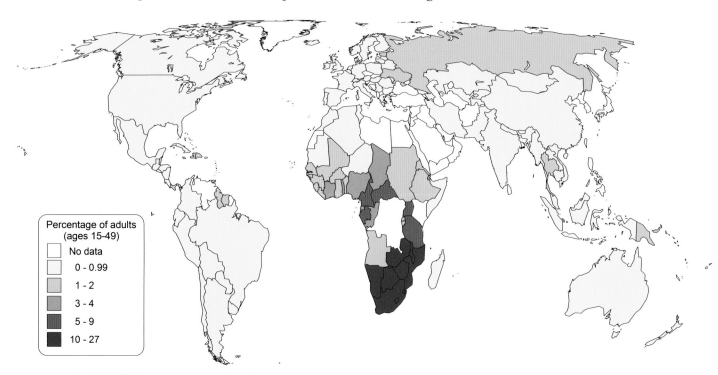

**Percentage of adults (ages 15-49)**

- No data
- 0 - 0.99
- 1 - 2
- 3 - 4
- 5 - 9
- 10 - 27

Map 29.1. HIV infections among adults, 2007.

**Table 29.1. HIV/AIDS prevalence rates for the most seriously affected countries (>5%), 2007**

| Country | Ages 15–49 (%) |
| --- | --- |
| Swaziland | 26.1 |
| Botswana | 23.9 |
| Lesotho | 23.2 |
| South Africa | 18.8 |
| Namibia | 15.3 |
| Zimbabwe | 15.3 |
| Zambia | 15.2 |
| Mozambique | 12.5 |
| Malawi | 11.9 |
| Central African Rep. | 6.3 |
| Tanzania | 6.2 |
| Gabon | 5.9 |
| Uganda | 5.4 |
| Cameroon | 5.1 |

Source: UNAIDS 2008

Figure 29.1. HIV/AIDS education sign along a road in rural Kenya.

Even as AIDS undermines food security, poor nutrition hastens the progression of HIV infection to full-blown AIDS and ultimately death. The HIV/AIDS disaster has significantly reduced life expectancy in the most severely affected countries. In Botswana life expectancy dropped by 30 years, from 65 years in 1988 to 35 years in 2004. In neighboring Zimbabwe, life expectancy fell from 56 years in 1982 to 37 in 2005.

The population pyramid for Botswana reveals the devastating impact that HIV/AIDS can have on the structure of the population (figure 29.2). The graph shows a dramatic decline in the number of active male and female workers in the 15–49 age groups as a result

of the AIDS epidemic. This restructuring of the population has led to a reconfiguration of rural households. In some of hardest hit areas, over half the households are now headed either by women (30%), grandparents (20%), or children (5%). Orphan children and elderly parents lack the vigor and necessary resources to farm, which results in reduced agricultural production. Premature deaths of men and women in the 30–50 age groups result in a loss of knowledge and skills that would normally be transmitted to the next generation. Whether this practical knowledge pertains to production, trade, or some service, its loss can reduce the resiliency of households and communities to future shocks to their livelihoods (FAO 2004a).

Research in a number of countries reveals that households afflicted by HIV/AIDS are initially resilient to shocks on household food supplies (UNAIDS 2006, 86). In Malawi, where 12% of adults have contracted HIV/AIDS, one study found that illness and death did not automatically result in food insecurity (Mtika 2001). Households drew on social support networks, especially the extended family, to obtain food, labor, and money to cope with shortages during periods when food stocks were low and labor needs were high. This reliance upon the larger community in which more fortunate individuals feel obligated to assist kin in dire situations has helped to cushion the effect of the AIDS pandemic. But social ties that bind individuals in need with extended family members can break under stress. There is a point at which some households find it difficult to share with others because of the strains it places on their limited resources. This breakpoint can be reached as a result of multiple pressures (poor harvests, low market prices, conflict) in addition to assisting an increasing number of HIV/AIDS-afflicted households. Reduced community-based assistance will increase the prospect of malnutrition, accelerating the health impacts of HIV/AIDS.

To avoid this deadly cycle of disease-malnutrition-disease, highly vulnerable communities require external as well as local assistance. To keep HIV from bringing hunger with it, agricultural development programs could emphasize less labor-demanding crops and technologies. Education and training programs could target orphans, single women, and the elderly, and food aid could focus on easily prepared and nutritious foods

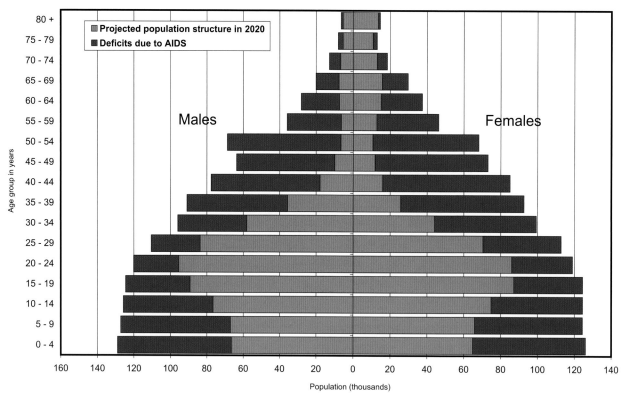

Figure 29.2. The projected demographic impact of HIV/AIDS in Botswana. *Source:* Ziegler and Ffrench 2000, fig. 4.

(FAO 2004a). Moreover, by making antiretroviral drugs affordable and easy to obtain, governments and pharmaceutical companies could enable people with HIV to work and lessen the effects of their illness on food insecurity. Supporting local economies and livelihoods requires a multipronged approach that includes im-proved access to clean water, health care, education, employment, and fair markets for agricultural inputs and products. Like many facets of the hunger problem, breaking the links between food insecurity and HIV/AIDS implies local and external initiatives.

# 30: Malaria

Malaria is a deadly disease transmitted by *Anopheles* mosquitoes. It is present in over 100 countries of the world and kills more than a million people each year. Over 90% of these deaths occur in Sub-Saharan Africa (map 30.1). Children under five years of age are malaria's biggest victims. In Africa, this group accounts for 90% of all malaria deaths, but adults suffer morbidity (illness) at high rates. The WHO estimates that 40% of the world's population is at risk of malaria. Tropical areas of the developing world face the greatest danger (WHO 2008f).

When malaria doesn't kill, it is debilitating. The disease is characterized by alternating fevers and chills and severe body ache. Those afflicted cannot work or attend school, resulting in reduced productivity, limited learning, and lost income. Malaria thus exacerbates poverty and hunger and induces other illnesses. For example, malaria is a major cause of anemia among children and pregnant women.

In 2006, 189–327 million people fell sick from malaria. An estimated 86% of malaria's victims lived in the WHO's Africa region (WHO 2008f). Another 28% resided in Southeast Asia with most of the cases in India. The poor are most affected by the disease. The Global Fund to Fight AIDS, Tuberculosis, and Malaria reports that 58% of the deaths caused by malaria take place among the poorest 20% of the world's population (Global Fund 2008). The costs of prevention and treatment are high for the poorest households. In Malawi, the costs of insecticide-treated sleeping nets and medicines to treat malaria amount to 20% of the annual income of the poor. The resistance of the main parasite (*Plasmodium falciparum*) to antimalarial drugs has increased the cost of treatment.

Some countries like Vietnam have successfully fought the disease through aggressive malaria-control programs that provide free antimalarial drugs and insecticides for treating mosquito nets. Working with

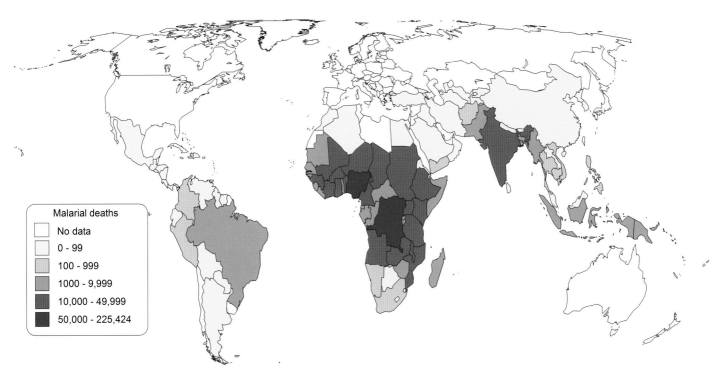

Map 30.1. Total deaths from malaria, 2006.

community-based youth and women's organizations has been key to Vietnam's success in reducing the malarial deaths from 4600 in 1991 to 148 in 2000.

The Global Fund provides grants in over 85 countries for malaria-control programs that focus on a new generation of antimalarial drugs and insecticide-treated bed nets. In December 2005, the Global Fund supported the distribution of two million insecticide-treated bed nets in Niger to mothers with children under the age of five.

# 31: Health Expenditures per Capita

Access to good health care can alleviate some nutritional problems for the poor, even under situations of low food availability. If health care coverage is thin, then food-deficient households are more vulnerable to health problems linked to hunger. Poor health negatively affects nutrition by reducing appetite, impeding the absorption of food nutrients, and competing with the body for energy that would normally go into growth and weight gain. Good health is critical to normal growth and development, especially for young children and pregnant women. One measure of the availability of health care is expenditure on health services per capita. Health expenditure per capita refers to the sum of private and public spending on health-related services divided by a country's population. Health services include preventive and curative treatment, nutrition programs, family planning, and health-related emergency aid.

According to the United Nations Development Program, health expenditures must amount to $30–40 per person to provide basic health care in developing countries (UNDP 2005, 63). The map of health expenditure per capita (map 31.1) shows that most of Sub-Saharan Africa (e.g., Mozambique, $14) and South Asia (e.g., Pakistan, $15) does not meet this level of spending. When we compare this pattern of poor health care to the geography of hunger (map 8.1), we see that most countries with high hunger vulnerability also have poor health coverage. This relationship stands out in figure 31.1, in which a dense clustering of countries is associated with low per capita health expenditures and high hunger vulnerability. There is noticeably less hunger vulnerability for countries spending more than $50 per capita. This combination presents a disturbing picture of extreme vulnerability among food-insecure households to health problems associated with malnutrition. However, this does not mean that low health spending causes hunger vulnerability. The same conditions of

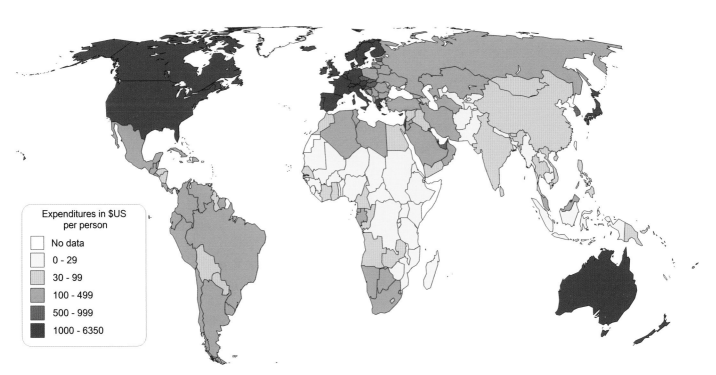

Map 31.1. Health expenditures per capita, 2005.

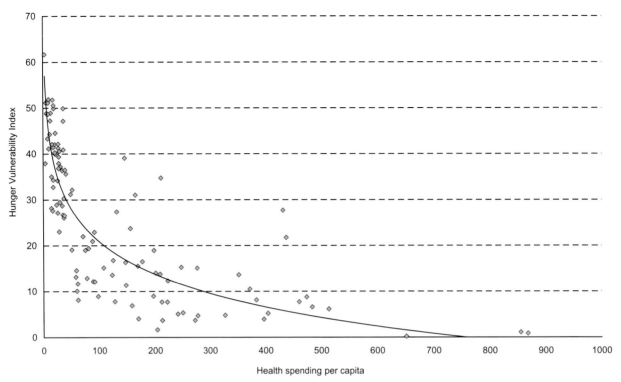

Figure 31.1. Health spending and hunger vulnerability. *Sources:* WHO 2008c; FAO 2008b; World Bank, PovcalNet, n.d.; World Bank, *World Development Indicators Online,* n.d.

poverty probably generate both poor levels of health care and hunger. The main lesson of map 31.1 and figure 31.1 is that hunger vulnerability is most prevalent in places where there is the least capacity to cope with the health problems that contribute to or are exacerbated by malnutrition.

At the country scale, map 31.1 reveals that health expenditures vary dramatically across the globe. The high-income industrialized countries of the global North show the highest health-spending levels. Latin America and the Caribbean and Southern and Northern Africa have moderate health-expenditure levels. The countries where the least amount of private and public income is devoted to health are found in intertropical Africa and South Asia.

The pattern of health inequalities apparent at the international scale is replicated at the subnational scale. Within any given country, including the United States, inequalities expose large segments of a country's population to nutrition-related diseases. Where governments do not provide universal health insurance and people must pay out of their own pockets for

basic health care, the poor suffer the most (figure 31.2). In Vietnam, a single hospital visit amounts to 40% of the monthly income of the poorest households. In the United States, 46 million people have no health insurance. Poor people and minorities are overrepresented in this disadvantaged group (Phelan and Link 2005).

Inequalities in health expenditures also exist between rural and urban areas within countries. In China, health expenditures are 3.5 times higher in cities versus rural areas. Prior to 1980, the Chinese government ensured basic health care coverage under its Cooperative Medical System. When that medical safety net was removed following market reforms in the late 1970s, health care coverage declined for a large segment of the rural population. Today, 70%–80% of China's rural population has no health insurance (UNDP 2005, 63). Yet rates of child growth failure and hunger vulnerability have declined in China. It is likely that higher individual incomes associated with economic reforms and industrialization have counteracted declining health care coverage—at least for households of higher socioeconomic status.

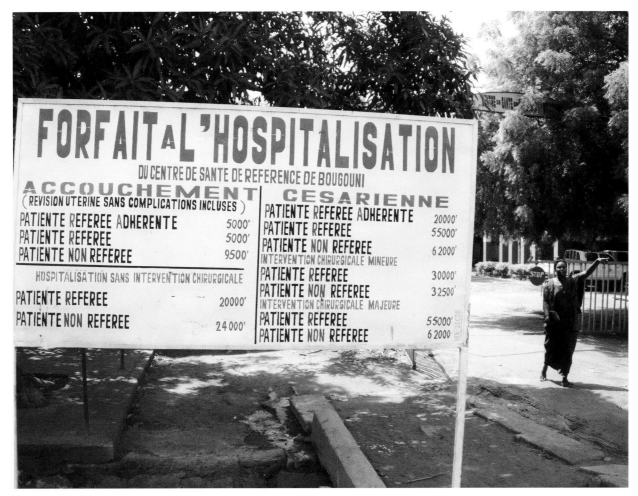

Figure 31.2. Prices for maternity clinic services posted outside the gates in Bougouni, Mali. The cost for delivering a baby can be as high as $20—or 20 days of agricultural wage labor for women. Poor households cannot afford these fees, which leads most women in rural areas to give birth at home with the assistance of village midwives.

# 32: Unsafe Water and Poor Sanitation

Hunger vulnerability is intensified by poor sanitation and limited access to safe drinking water. Water-borne diseases such as diarrhea, dysentery, and cholera afflict the poorest households in the developing world. Diarrhea alone kills 2.2 million people each year. Its victims are mainly children under the age of five (UNICEF 2007). These diseases occur when untreated sewage flows into the sources of people's drinking water. Malnutrition lowers a person's resistance to such diseases, intensifying their effects. At the same time, these diseases sap the body of essential nutrients, which exacerbates the impact of poor diet. Improved water supply and sanitation save lives and improve nutrition.

Target 10 of the UN Millennium Development Goals is to cut in half the proportion of people who do not have access to safe drinking water and basic sanitation by 2015. To reach this goal, considerable commitments need to be made by both aid donors and governments. Good progress was made between 1990 and 2004, when 86 million more people benefited each year from improved sanitation. To attain the 2015 target, that number must increase to over 135 million per year.

The WHO-UNICEF joint monitoring program uses household surveys and national censuses to determine the proportion of the population with access to improved water sources and sanitation facilities. It periodically updates its website (www.wssinfo.org) with revised coverage estimates. The WHO measures adequacy of sanitation as the percentage of the population that uses improved sanitation facilities. Improved facilities include connection to a public sewer, a septic system, a pour-flush latrine, a simple pit-latrine, or a ventilated improved pit latrine. Access to safe drinking water refers to the percentage of the population that

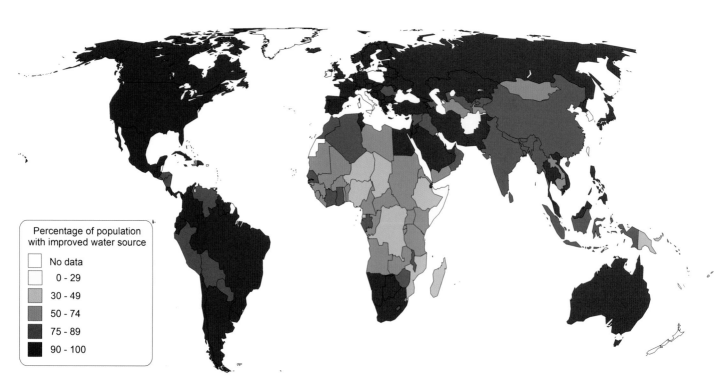

Percentage of population with improved water source

No data
0 - 29
30 - 49
50 - 74
75 - 89
90 - 100

Map 32.1. Improved drinking water, 2006.

has 20 liters (about 5 gallons) of water per person per day from an improved water source. Improved sources include a household connection, a public standpipe, a borehole, a protected dug well or protected spring, and rainwater collection. The WHO-UNICEF data do not indicate the quantity of safe water obtained from these sources. As a result, these data may overestimate access to safe drinking water (McKenzie and Ray 2005, 4).

In 2006 about one out of six people in the world lacked access to clean water. At the same time 41% of the world's population went without adequate sanitation facilities. The majority of these disadvantaged and vulnerable people live in South-Central Asia, Eastern Asia, and Sub-Saharan Africa. As maps 32.1 and 32.2 show, less than half of the people in these regions have access to clean water and sanitation. These are the same regions where hunger vulnerability is greatest (map 8.1).

On the one hand, one can point to poverty as the underlying condition that helps to explain the coincidence of hunger vulnerability and limited access to improved water and sanitation systems. However, as the case of Angola indicates, some governments have the financial resources to ensure for the provision of safe water and sanitation, but lack the political will and priorities to address the issue. At $2360 (PPP), Angola has roughly the same per capita income as Ghana, Bangladesh, and Pakistan. But while 75% to 90% of the population in those three countries has access to clean water, only about half of Angolans enjoy that basic service.

A *New York Times* front-page feature story on June 16, 2006, revealed that filthy water and politics go hand in hand in Angola (LaFraniere 2006). In the capital city of Luanda, sewage flows into the Bengo River, a major source of water for the city's poorer residents. The more affluent have clean water piped into their homes. The majority of citizens buy water for 12¢ a gallon from some 10,000 private vendors who fill their tanks daily with untreated river water. This system works for some people. Indeed, according to UNICEF, access to safe water in urban Angola rose from 23% of the population in 1990 to 75% in 2004. But the poor are neglected with fatal results. Between February and June 2006, cholera struck the city and the countryside, claiming 1600 lives. Another 46,000 fell sick from drinking, bathing, washing clothes, and cleaning with contaminated

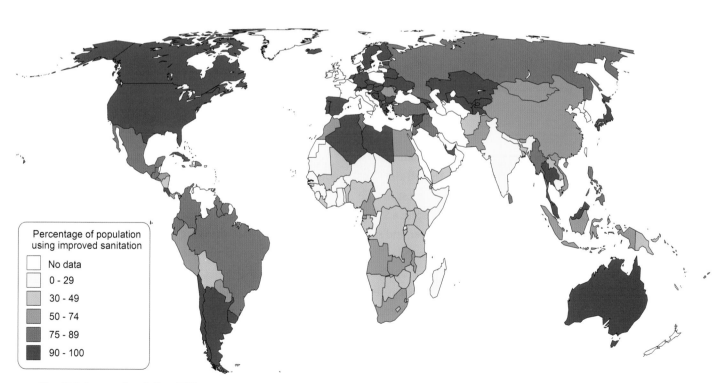

Map 32.2. Improved sanitation, 2006.

Percentage of population using improved sanitation

- No data
- 0 - 29
- 30 - 49
- 50 - 74
- 75 - 89
- 90 - 100

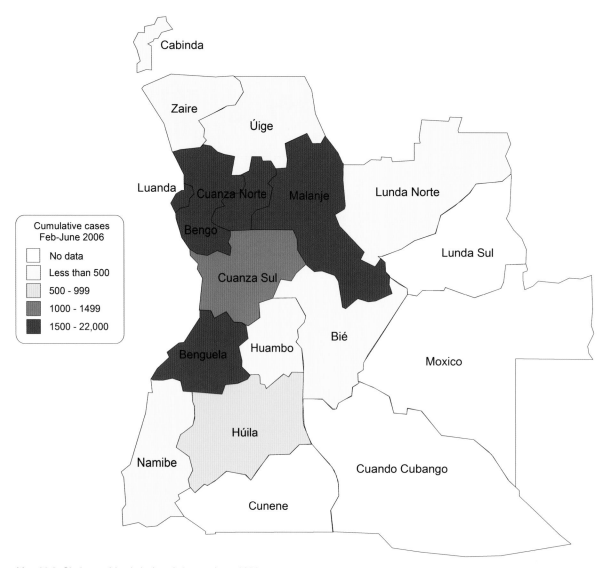

Map 32.3. Cholera epidemic in Angola by province, 2006.

water (map 32.3). Angolans living on less than $2.00 a day could not afford clean water, but the oil-rich Angolan government had the resources to invest in safe drinking water and sanitation facilities. Despite a $2 billion budget surplus in 2004, improving water and sanitation was not high on the government's agenda. Even the government's response to the cholera epidemic was slow and underfinanced. Whether it is because of the government's inability to get back on its feet after almost three decades of civil war or massive corruption that siphons off public money into private pockets, Angola's terrible water and sanitation situation leaves its poor to live and die in misery.

Scarcity and abundance are legendary in India, a country that has exported millions of tons of food grains while millions of people go hungry. Millions also suffer from water-transmitted diseases because of poor access to safe drinking water. The combination of malnutrition and disease is deadly for millions of infants and preschool children each year (Swaminathan Research Foundation and World Food Program 2001, 157–58). India's National Family Health Survey for 2005–6 indicated that 95% of urban and 85% of rural households enjoyed access to safe drinking water (IIPS and Macro International 2007). This coverage was up from 1991, when just 81% of urban and 56%

Figure 32.1. Drawing water from a passive, traditional water harvesting system (*kund*) in rural Rajasthan, India. Photo by Kathleen O'Reilly.

of rural populations had sufficient access (McKenzie and Ray 2005, 4). There were, however, significant inequalities in access to improved water among states. Access to safe drinking water surpassed 95% in three states (Bihar, Haryana, and Punjab). However, a third of the households in four states drew water from unimproved sources (Manipur, Jharkhand, Meghalaya, and Nagaland (IIPS and Macro International 2007, 42)

(map 32.4). The data show a strong relationship between the wealth of a state and access to safe water (McKenzie and Ray 2005, 4).

Access to improved water quality varies geographically between rural and urban areas as much as among countries and social classes. Figure 32.2 reveals that urban populations are more likely to drink clean water than rural populations. In comparison to 1990, the percentage of people drinking safe water has increased. Between 1990 and 2004, worldwide, coverage rose from 78% to 83%; for the least developed countries, it rose from 51% to 59%. The cases of Bolivia (see box 32.1), India, and Angola remind us that even within cities of the world, striking inequalities exist among social groups over access to safe drinking water. Clean water is important for good nutrition. One of the challenges in ending hunger is to make this coupled public health problem a political priority.

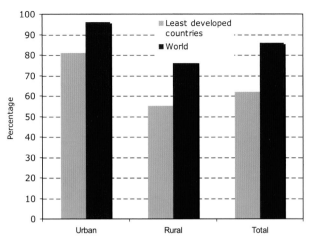

Figure 32.2. Percentage of population with access to improved water, 2006. *Source:* World Bank, *World Development Indicators Online,* n.d.

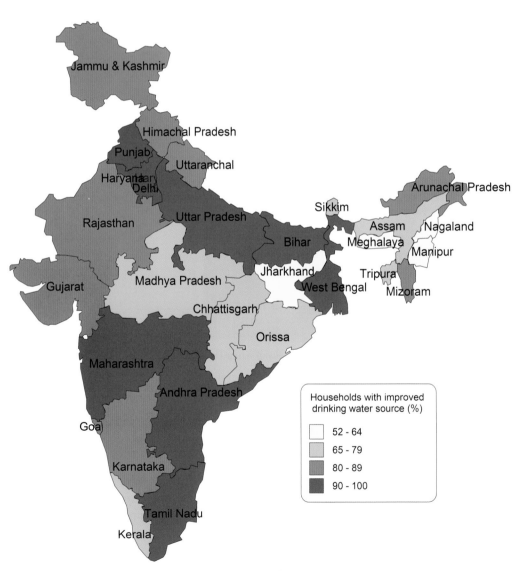

Map 32.4. Indian households with improved drinking water, 2005–6.

**Box 32.1. Water Privatization Revolts in Bolivia**

How to close the clean water gap between rich and poor, between rural and urban areas, and between states in countries like India and Angola is a major policy concern. During the periods described in this text, urban water was provided by public agencies in Angola and India. Proponents of market-based approaches argue that the private sector can more efficiently expand water and sanitation networks to address the needs of underserved groups and regions. Under pressure from the World Bank to reduce government spending in the context of a structural adjustment program, the Bolivian government signed water privatization contracts with transnational water companies in the late 1990s. The French firm Suez took over the formerly public water system of La Paz in 1997. International Water, controlled by the giant US engineering firm Bechtel, won a 40-year contract and a guaranteed 15% annual profit rate to control Cochabamba's water system in 1999. In Cochabamba, after water rates doubled overnight for poor households , a broad coalition of rural and urban water users protested the sale of what they considered to be a public good and a human right that should not be owned by a profit-making enterprise. The broad social movement that took to the streets in Cochabamba and forced the government to reverse its decision in 2002 is now famous in the antiglobalization literature. A second revolt against water privatization took place in La Paz in 2005 in the poor Indian neighborhood of El Alto, where tens of thousands of people had no access to improved water despite the promises of Suez and the World Bank. The protests forced the Bolivian government to terminate Suez's contract and transfer control of La Paz's water and sanitation systems to a public agency that, in turn, did little to reduce water rates and expand services. In response to ongoing protests in La Paz, the new government of Eva Morales appointed the leaders of the Cochabamba and La Paz water revolts to head a new Water Ministry whose priority is to make improved water and sanitation accessible to the poor and underserved (Lobina 2000; Finnegan 2002; Shultz 2005).

# 33: International Trade and Primary Products

There are many national and international economic conditions that can expose people to greater risk of hunger. Maps 33.1–38.4 explore how a country's position in international markets and its access to development assistance relate to hunger.

The role of international trade in world hunger stirs heated debates. On one side sit those who argue that international trade provides new opportunities for economic growth and poverty reduction in developing countries. This positive view focuses on instances when international markets create jobs for the poor in cities like Dhaka, Bangladesh, or new income sources for impoverished farmers in places like rural Guatemala. On the other side sit those who see the nature of international trade as inherently unfair. They report frequent cases in which increased trade between rich and poor countries has brought greater inequality in countries like Mexico and China. As box 33.1 indicates, there is evidence to support both views.

While disagreement persists about the promises and pitfalls of international trade, there is agreement that countries that depend on the export of primary products such as unprocessed minerals or agricultural goods are disadvantaged. Concentration in primary commodities contributes little to other parts of a country's economy, like manufacturing and services. Moreover, countries exporting raw materials are vulnerable to dramatic losses in export earnings if prices fall or crops fail. Map 33.1 shows the share of a country's total exports that are primary products. Countries that depend on a few primary products like coffee and cotton for a large percentage of their national income are likely to have low levels of income and are also likely to experience shocks to their income (figure 33.1). These conditions of high export concentration and income instability are conducive to greater poverty and hunger.

Map 33.2 indicates the intensity of trade in a country measured by the sum of the country's imports and

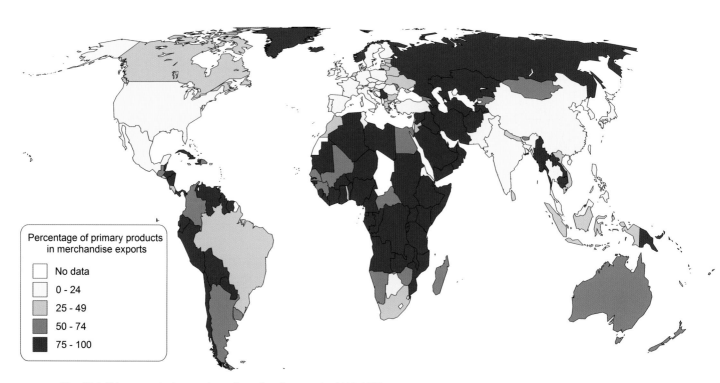

Percentage of primary products in merchandise exports

- No data
- 0 - 24
- 25 - 49
- 50 - 74
- 75 - 100

Map 33.1. Primary products as a share of merchandise exports, 1999–2003.

**Box 33.1. International Trade, Poverty, and Inequality**

In the 1990s, increased access to international markets fueled growth in Bangladesh's garments exports, which led to the creation of almost 2 million jobs, mostly for poor women. Increased incomes in the garments sector have been linked to reduced poverty and improved health and education. (Bhattacharya 2003, in UNDP 2005). On the other side of the globe, in the rural highlands of Guatemala, over 80% of the population lives in extreme poverty, but in the 1990s about 21,500 families were able to gain access to the US market for snow peas and broccoli. The exports brought in over $30 million a year, which was spread over the thousands of smallholders and contributed to the alleviation of poverty and hunger for over 150,000 people (Gulliver 2001 in Dixon, Gulliver and Gibbon 2001).

International trade creates opportunities, but it can also expose global inequality. As a result of women flocking to jobs in garments factories in South and Southeast Asia, consumers in the rich world can purchase lower-priced clothes. The contrast between the harsh and often dangerous conditions of factories in poor countries and the comfortable lives of those who buy the final products reveals the global inequalities that exist (Rivoli 2005). While recent growth in international trade has alleviated poverty for many people, its benefits have often offered more to the rich than to the poor. During the 1990s, the volume of world trade almost doubled, and the benefits of growing trade seemed to be skewed toward richer countries and toward the wealthier within countries. Some of the poorest countries in the world, mostly in Africa, have been largely left out of the expansion in trade, while countries that are increasingly active in international trade, like Mexico and China, experienced marked increases in pay inequality among workers (Galbraith 2004).

Figure 33.1. Bales of cotton stacked outside a gin in Houndé, Burkina Faso. Ginning removes the seed in cotton bolls and cleans the cotton lint before it is baled. Burkina Faso is a developing country in West Africa that is dependent on the export of primary products. In the three-year period 2001–3, cotton exports alone accounted for 36% of the country's foreign exchange earnings. Because there is no domestic textile industry, all of Burkina's cotton is exported once it leaves the gin.

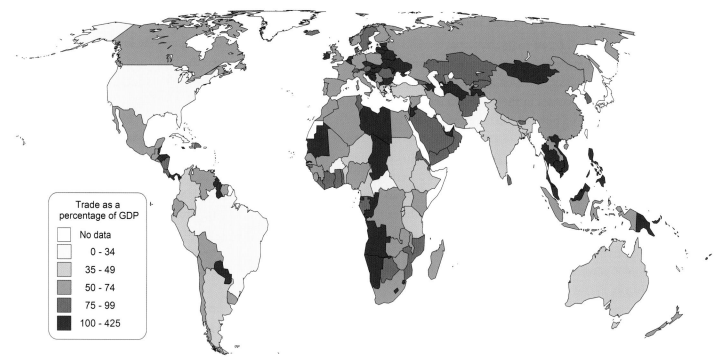

Map 33.2. Trade (exports and imports) as a share of GDP, 1997–2002.

exports as a share of its national income. Many developed countries, like the United States, engage in a great deal of international trade, but trade remains a small share of their total economy because the domestic market is so large. Other high-income countries that are fairly small, like Holland, rely on international trade for a large share of their national income. Developing countries also tend to have a high share of their incomes based on exports, but these countries usually rely much more on primary products. Such goods comprise a small share of European trade, but typically account for three-quarters of Africa's exports and over half of Latin America's.

Heavy reliance on international trade combined with a concentration of exports in primary products describes many of the regions that experience high levels of hunger vulnerability (Sub-Saharan Africa, Central America and Andean South America, Southeast Asia). India is an exception to this tendency in that it has high hunger rates but relatively diversified exports. The shift from the export of primary products to other goods and services is a recent phenomenon in India.

# 34: International Terms of Trade

A likely causal link between trade in primary products and hunger is through prices. If a country depends heavily on the export of a few goods and the price of those goods on the world market falls, the country will lose income and experience greater poverty. It will also find it difficult to pay for vital imports such as food to eat and fuel to run its industries. A similar problem arises if there is an increase in the prices of goods that a country imports. It becomes increasingly difficult to pay for these imports if they become more expensive compared to the value of the country's exports. A fall in the price of exports or an increase in the price of imports can therefore contribute to poverty and hunger.

The terms-of-trade index is a measure of the prices a country receives for its exports in comparison to the prices it pays for its imports. It is measured as the price of exports divided by the price of imports. An increase in the price of exports or a decrease in the price of imports results in increasing terms of trade. Conversely, a fall in the price of exports or a rise in the price of imports corresponds to falling terms of trade. When a country experiences rising terms of trade it can import more goods and services for the same volume of exports. If terms of trade are falling, the country can afford fewer imports for the same quantity of exports. The result can be greater poverty and hunger.

Terms of trade could play an important role in the hunger problem through either long-term trends or through sudden changes. Map 34.1 indicates the average change in terms of trade over the period 2000–2006. Most developed economies enjoy fairly stable terms of trade (though over this period they were often declining slightly). The relative stability emerges partly because these countries have highly diversified exports. When export prices for one commodity fall, production and exports shift to other commodities, so that average export prices are stabilized. Less developed countries that are dependent on primary products are more likely to

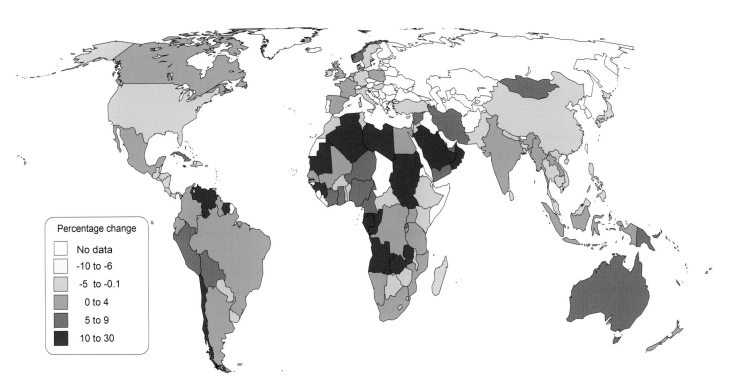

Map 34.1. Average annual change in terms of trade, 2000–2006.

**Box 34.1. World Cotton Prices and Hunger Vulnerability in West and Central Africa**

The link between a country's dependence on primary goods and hunger vulnerability is clear in the case of cotton. For some 2 million farmers in West and Central Africa, cotton is the principal cash crop. Not only is it the main source of income for rural households, but governments also depend on cotton exports for a large percentage of their foreign exchange earnings. Cotton accounts for more than a third of Burkina Faso's export earnings and nearly one-fifth of Mali's. However, the value of cotton exports dropped precipitously in the late 1990s and early 2000s when the world market price fell to 42 cents per pound. After adjusting for inflation, world cotton prices in 2001–2 were at their lowest levels since the 1930s Great Depression. Many analysts were pessimistic that the price would not go above 50–60 cents per pound in the next 10 years, well below the long-term average of 72 cents (Oxfam 2002).

Many studies attributed the price shock to subsidies provided by European and especially the US government to domestic cotton growers (FAO 2004b). US farmers are unable to produce cotton for less than 70 cents per pound. To make up for the difference between their production costs and market prices, they lobby the US government for subsidies. Thanks to their successful lobbying efforts, the average payment received by cotton producers in the early 2000s amounted to $120,000 (World Bank 2008b, 99). As a result of this support, US cotton growers continued to produce cotton at near record levels despite historically low market prices. Since US cotton accounts for 40% of world exports, high cotton production in the United States ends up depressing world market prices. The World Bank estimated that world cotton prices fell by 10%–15% because of US agricultural subsidies.

Rich-country agricultural subsidies are not the only force producing poverty in the rural economies of developing countries. Because cotton is sold in world markets in US dollars, the exchange rate between national currencies and the US dollar also influences cotton grower incomes. A weak US dollar and a strong euro, to which West and Central African currencies are linked, resulted in African farmers receiving less local money for their exports. An example of the effect of currency exchange rates on producer incomes stood out in the period August 2002–January 2004. World cotton prices rose by 54% over that period, but due to unfavorable exchange rates, the price increase in national currencies amounted to just 19% (Estur 2005).

Falling incomes from agricultural subsidies and unfavorable exchange rates increase poverty and hunger vulnerability in cotton-growing households of West and Central Africa (Minot and Daniels 2005). Farmers depend on cotton earnings to invest in new technologies that benefit food crops as well as cotton. When cotton incomes decline, the production of food staples like millet, maize, and sorghum also declines (OECD 2005). François Traoré, the head of the West African Cotton Producers' Association, pointed to the heightened degree of hunger vulnerability under these conditions when he told a BBC reporter, "When [the cotton price] falls, one cannot look after the family" (BBC 2006).

experience periods of rapidly increasing or decreasing terms of trade. Over the period 2000–2006, petroleum exporters such as Nigeria, Libya, and Venezuela experienced large increases in their terms of trade, while many exporters of agricultural products, like Kenya, and petroleum importers, like China, suffered persistently declining terms of trade.

Persistent declines in terms of trade exacerbate poverty and hunger in low-income countries. Because terms of trade are just one of many factors influencing hunger, there is little relationship between hunger vulnerability scores and the terms-of-trade map. Countries with declining terms of trade include Thailand, with fairly low hunger vulnerability, and Nicaragua and Ethiopia, where hunger vulnerability is high. Countries with improving terms of trade include Nigeria, with very high hunger vulnerability, and Algeria, where hunger vulnerability is low. While many factors influence the hunger outcomes, it is fair to say that alleviating hunger is more difficult when terms of trade are in decline.

Even where terms of trade have been increasing over time on average, it is possible that large, temporary declines in terms of trade could force many people into hunger. The possibility that a sudden terms-of-trade shock can exacerbate hunger is explored in maps 35.1 and 36.1.

Figure 34.1. Francois Traoré, president of the West African Cotton Producers' Association, discusses unfair trade practices in world cotton markets at an APROCA meeting held in Orodara, Burkina Faso, June 22–23, 2005. The goal of the organization is to defend the interests of West African cotton growers through concerted actions in such venues as the World Trade Organization.

# 35: Terms-of-Trade Shocks

Regardless of the long-term trends in prices, sudden declines in terms of trade can have an immediate effect of worsening conditions for the poor. These shocks in terms of trade are especially frequent in countries that rely heavily on the export of a few primary products.

The terms of trade will decline if the price of a country's primary export, say coffee, falls, or the price of an important import, say oil, rises, or if both happen at the same time. A terms-of-trade shock can exacerbate food insecurity by reducing the incomes the poor earn from producing goods for export or by increasing their costs of consuming imported goods. Terms-of-trade shocks are much more common in poor countries than in rich ones (map 35.1). While affluent economies tend to trade in a wide range of goods and services, exports from poor countries are often highly concentrated on very few products (see map 33.1). As a result, a decline in the price of a single good can cause a substantial terms-of-trade crisis. An example of this kind of shock

is presented in box 35.1, describing the coffee crisis that destabilized economies in Latin America and Africa. As can be seen in map 35.1, most of the countries that experienced a terms-of-trade shock of over 5% (a 5% decline in the price of exports compared to the price of imports) in any single year are either African economies that export only a few minerals or agricultural products, or petroleum exporters.

There is some correspondence between the countries that experience terms-of-trade shocks and those that have higher rates of hunger. This correspondence emerges partly because poor countries tend to be at risk of both hunger and export price shocks. As figure 35.1 demonstrates, terms-of-trade shocks appear to exacerbate hunger in many countries, but they have particular impact in countries with high rates of poverty. This figure plots the average rates of child growth failure for countries with different poverty rates and different terms-of-trade experiences. Comparing those

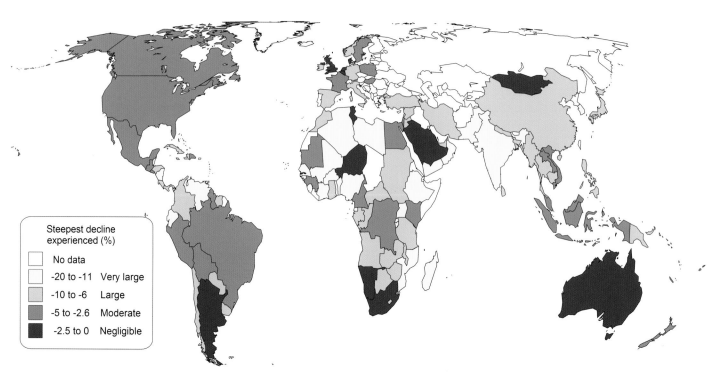

Steepest decline experienced (%)

| | |
|---|---|
| No data | |
| -20 to -11 | Very large |
| -10 to -6 | Large |
| -5 to -2.6 | Moderate |
| -2.5 to 0 | Negligible |

Map 35.1. Largest single-year decline in terms of trade, 2000–2006.

**Box 35.1. The Coffee Crisis**

No matter where a cup of coffee is drunk, the chances are that the beans that went into it were grown on a small farm in a poor country. Around the world, about 25 million people grow coffee and about three-quarters of that coffee comes from small family farms of less than 20 acres. In Brazil, where coffee is often grown on large plantations, 3 million workers in the coffee industry depend on the crop for their wages. Many countries rely on coffee as the primary national export and a major employer of the rural population. For example, Burundi earned over three-fourths of its export revenue from coffee in the year 2000 while coffee made up over 40% of the exports of Ethiopia and Uganda. In terms of employment, one out of four Ugandans and one out of ten Hondurans earned their income from coffee cultivation. Because so many people rely so heavily on coffee, a recent terms-of-trade shock against the crop was justifiably described as "the coffee crisis."

Coffee prices began to fall precipitously toward the end of the year 2000, dropping to less than one-third of their 1997 peak by the end of 2001. Those prices showed scant recovery in the four years after that. Adjusting for inflation, prices remained far lower than ever before, less than 25% of what they had been in the 1960s and 1970s. The dramatic drop in prices for coffee exports in 2000 was caused by rapidly increasing coffee production, especially in Vietnam and Brazil, and stagnant consumption. The oversupply created a glut of coffee beans and low prices (Gresser and Tickell 2002; Ponte 2002).

For many small-scale farmers and laborers, the coffee crisis meant intolerable poverty. In Central America, 600,000 coffee workers lost their jobs as a result of the crisis. In Africa, coffee grower incomes fell to near zero. In March 2002, the World Food Programme estimated that the crisis was so severe that it pushed 30,000 Hondurans into hunger. Agencies throughout Central America reported heightened malnutrition in coffee growing areas as the crisis deepened. Across the world, the coffee crisis meant that sick children could not have medicine, school fees could not be paid, and hunger could not be avoided. In response to the crisis, governmental and nongovernmental organizations sought to restructure world coffee markets through price stabilization plans. For their part, socially conscious consumers began buying "fair-trade" coffee that is certified by organizations like TransFair USA that ensure that farmers and farmworkers are paid a fair or above-market price for their products and labor (Jaffee 2007). However, demand for fair-trade coffee is not robust: just 1%–2% of the coffee produced in the world is sold at fair-trade prices (FLO 2007).[1]

1. Fair-trade coffee sales amounted to 66,219 tons in 2007, compared to total production of 7.7 million tons and total exports of 5.5 million tons.

countries that experienced a single-year decline in their terms of trade of over 10% with those that did not, we see that rates of child growth failure are about 7 percentage points lower when shocks were absent (29% are opposed to 36%). Among countries with rates of $2.00/day poverty of over 50%, the prevalence of growth failure is greater and the impact of shocks seems to be larger. Those highly impoverished countries that experienced terms-of-trade shocks have rates of growth failure of 50% compared to 39% for similarly poor countries that avoided large shocks. Countries with lower rates of poverty have lower rates of growth failure and show a somewhat less dramatic difference between those that experienced trade shocks and those that did not.

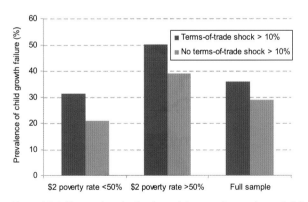

Figure 35.1. Terms-of-trade shocks and the prevalence of growth failure. *Sources:* WHO 2008c; FAO 2008b; World Bank, PovcalNet, n.d.; World Bank, *World Development Indicators Online*, n.d.

# 36: Food Trade

How does a country's imports or exports of food relate to its people's vulnerability to hunger? Not all countries have the resources to provide their populations with the foods they want from domestic production. Some countries, like Japan, have large populations and little capacity to produce, while others, like Australia, can grow far more food than their small populations require. This imbalance between capacity to produce food and need to consume it is reflected in patterns of international trade in cereals.

Map 36.1 shows net trade in cereals for the countries of the world. The data are calculated by the FAO as the value of exports of maize, wheat, rice, barley, and other grains minus the value of imports of the same cereals. Countries that import more than they export (net food importers) have negative net trade, while those that export more than they import (net food exporters) have a positive net food trade. In 2004, there were about 30 net food exporters in the world and 150 net food importers.

Whether a country is a net importer or exporter of food has little bearing on its hunger problem. In 2004, the largest food importer in the world was Japan, with net cereals imports of over US$5 billion. Meanwhile the largest food exporter was the United States, with net cereals exports of over US$12 billion. While hunger exists in both of these countries, neither of them faces a widespread undernourishment problem. Many of the large food importers (those with over $1 billion worth of net cereals imports) are high-income countries that import grains for animal feed to support meat consumption by affluent consumers. Major food exporters (those with net cereals trade in excess of $1 billion) are almost all high- or middle-income countries with large quantities of arable land relative to their populations.

It is tempting to see a growth in food imports as a troubling sign for a country's food security, but the relationship between food trade and hunger is not so direct. Growth in incomes, as has recently been experienced in China, can lead to rapid increases in food demand and

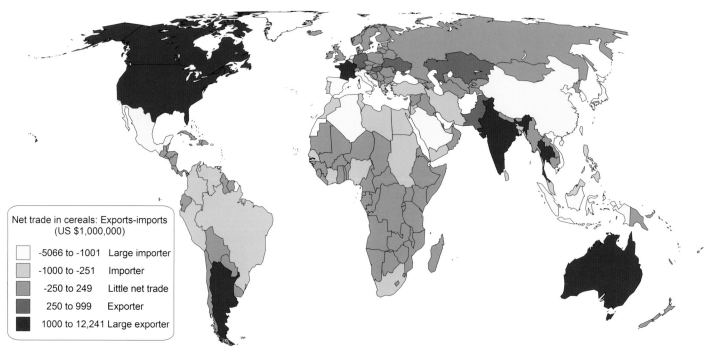

Net trade in cereals: Exports-imports
(US $1,000,000)

| | |
|---|---|
| -5066 to -1001 | Large importer |
| -1000 to -251 | Importer |
| -250 to 249 | Little net trade |
| 250 to 999 | Exporter |
| 1000 to 12,241 | Large exporter |

Map 36.1. Cereal trade, 2004.

**Box 36.1. Biofuels, Global Food Availability, and Prices**

Since the year 2000 there has been growing enthusiasm in Europe and North America for using renewable fuels based on agricultural products. Corn-based ethanol and biodiesel from vegetable oils are widely embraced as tools for reducing reliance on imported petroleum while reducing the negative environmental effects of burning petroleum-based fuels. Whether or not these biofuels contribute effectively to energy independence or environmental protection, their expansion has had a tremendous effect on global food availability and food prices. This impact has been negative for many of the world's poor.

In 2001, the United States produced about 1.7 billion gallons of ethanol using about 18 million tons of corn. By 2007 ethanol production exceeded 6 billion gallons and required 80 million tons of corn (Renewable Fuels Association, 2007; FAO, 2007b). In other countries there were similar expansions in biodiesel production, which implied huge growth in demand for palm oil and oilseeds. The increase in nonfood demand for cereals and other crops contributed to unprecedented increases in food prices in 2007. That year saw a 25% increase in the use of foods for fuel, a smaller increase in use of cereals in animal feed, and a serious drop in crop yields in some production areas (such as Australia). Overall, 2007 saw record-high cereals production (2101 million tons), but consumption of cereals exceeded production, as it had in the previous five years as well. Global reserves of food were drawn low and cereal prices in international markets soared by almost 40% (FAO 2007a; von Braun 2007). Prices for specific crops rose even more. The price for palm oil, an important cooking oil in much of the world, rose 70%. Rice prices doubled in the next year. Higher prices in international markets mean that food prices rise in local markets. Countries that must import have to pay more for

those imports. Unless the government has funds for food stamps or similar programs, consumers will experience higher prices. At the same time, producers in countries that have surplus crops will have the opportunity to export them at high prices. Local consumers will have to match those high prices if they are to secure the locally produced food. In some cases governments will ban exports and set price limits, but such interventions are difficult to enforce and often prove counterproductive (see box 37.1). At a national level, countries that export cereals benefit from these high prices, which raise their terms of trade, while importers suffer (see maps 34.1 and 36.1). At the household level, impoverished people who do not grow all the food they eat will be at greater risk of hunger. Faced with higher prices, the poor are forced to eat less and to drop nutrient-rich foods from their diets in favor of staple grains.

By increasing nonfood demand for cereals, the biofuels boom has altered the world's food balance sheet. As described above, reduced availability of cereals for human consumption has triggered higher prices that effectively reduce people's access to food. By linking food to fuels, biofuels also affect variability in food prices. Petroleum and energy prices have been highly volatile. Because a growing share of cereals demand is now for fuel production, cereals prices now move with changes in petroleum prices. This implies another problem for the poor and for governments trying to assist them. Not only are food prices likely to be higher for some time, food price shocks are also likely to become more frequent. In October 2008, the United Nations called on developed countries to "urgently" review their biofuel subsidies and policies because of their impact on rising food prices and hunger in the developing world (Rosenthal 2007).

food imports even as people's nutrition improves. At the same time, food exports do not imply that hunger is diminishing. India regularly exports huge quantities of grain (map 36.1) while millions of its citizens go hungry. Domestic production soared during the second half of the twentieth century thanks to new technologies and government subsidies to grain farmers. While India's domestic production is exported, rates of malnutrition in the country remain high because domestic consumers are too poor to buy the food their bodies need. In

contrast, rising imports into China reflect reductions in poverty that have enabled people to increase their consumption of a variety of foods.

Map 36.1 shows that the great majority of countries are modest food importers. This is partly because few countries have large enough populations and high enough income to import great volumes of food. On a per capita basis net food imports are also modest. Of the 150 importers mapped, 90 had net imports of under US$20 per person in 2004. Only 18 imported more that

Figure 36.1. Sacks of imported and domestic rice for sale in Abidjan, Côte d'Ivoire.

$50 per person per year. All of the large per capita importers were high-income countries. Among the other importers there was a mix of rich and poor countries with all manner of hunger conditions represented.

Regardless of how much a country imports or exports, the local food economy will likely be affected by global food markets. Large increases in global demand will drive up prices for consumers, whether their country tends to import or export. The intense demand in 2007 for corn and oil crops to produce biofuels in the United States and Europe led to price spikes for these staples in all countries (box 36.1). One indicator of the vulnerability of the poor to rising food prices on world markets was the large number of "food riots" that took place that year and in 2008. Public demonstrations against soaring prices for staple foodstuffs took place in Mexico, Morocco, Mauritania, Senegal, Guinea, Uzbekistan, and Yemen (Bradsher 2008).

Food trade is a necessary tool for moving grain from where it can be grown to where consumers are found (figure 36.1). But patterns of cereals trade say very little about where hunger is located.

# 37: Food Price Shocks

When people are vulnerable, any number of shocks can easily push them into crisis. Shocks can stem from a natural disaster or a violent conflict, but they can also result from a sudden change in market conditions that undermines people's welfare. For affluent people, these market shocks can be inconvenient or even disruptive, but they are rarely life-changing. For people living in poverty and lacking public or private assistance, a sudden shift in market prices can push them into hunger.

Food price shocks are the most obvious market disturbance that results in increased hunger. When local food supplies are low due to drought, pest invasion, or war, prices usually rise, often to levels that make food unaffordable for the poor. Similarly, food prices may suddenly rise if there is a change in international markets, which happens when a large country or group of countries unexpectedly imports more than usual or fails to export. Recent growth in demand for US corn by ethanol producers, for example, resulted in reduced corn for exports. Meanwhile, rising incomes in East Asia have created growing demand for meat in people's diets and growing imports of corn for animal feeds. These and other factors pushed up the prices that consumers must pay for corn. In Mexico those consumers responded with the "Tortilla Protests" of 2007 and 2008. Unprecedented increases in prices of other cereals stimulated similar protests around the world.

Food price increases impose a greater burden on the poor than the affluent, because poor people tend to spend a large share of their budgets on food. A household living on $1.25 per person per day may spend 75% of its income on food, while a typical household in a rich country will spend only 15% of its income in that way. This means that a doubling of food prices would increase the total food expenses of the poor household by another 75%, to a total of 150% of income, but it would raise the costs to the rich household by only 15%, to a total of 30% of income.

Map 37.1 shows the increase in food prices that countries experienced over the years 2000 to 2004, before the recent bout of global food price inflation. In some cases, rapid food price increases were in the context of general inflation and economic collapse, but in others the inflation was more concentrated in food. In either case, when prices rise and the incomes and entitlements of the poor do not, the threat of hunger intensifies. Over the five-year period covered in this map, food prices more than doubled in 10 countries. In some countries they doubled in a single year. In Belarus, food prices more than tripled in 1999 when the country's economy entered a deep crisis. Crop failures and restrictions on imports and aid distribution led to a doubling of food prices in Zimbabwe in 2002. General inflation caused food prices to explode thereafter.

The impact of this kind of food price inflation on hunger will vary significantly with the general poverty and vulnerability of a population. In an affluent region, price shocks are less likely to result in widespread hunger than in a poor one. Both Romania and Malawi experienced food price increases of over 40% in a single year. About 90% of Malawi's population lives on less than $2.00 a day compared to under 5% in Romania. Given the higher levels of poverty and hunger vulnerability in Malawi, we find more hunger there than in Romania.

Because the impact of price shocks on hunger varies with a country's poverty level, the map of food inflation (map 37.1) does not show much correlation with the hunger vulnerability map (map 8.1). However, if one controls for income level by looking only at countries with high poverty rates, then the impact of price shocks is more apparent. Figure 37.1 divides the countries of the world into two groups: those with very high poverty rates (over 50%) and those with lower poverty rates (less than 50%). For each of these groups, the figure plots the average prevalence of child growth failure for countries that experienced food price shocks and those that did not. A shock is defined as occurring

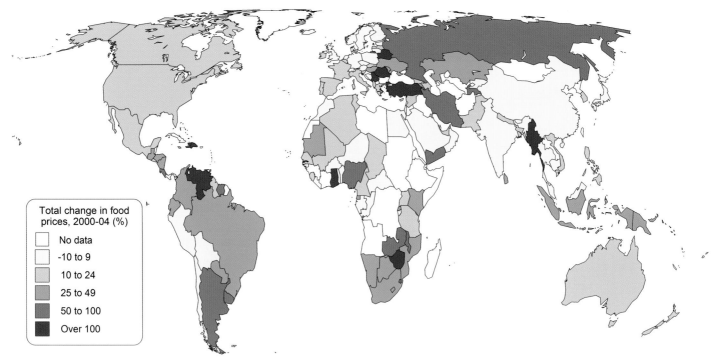

Map 37.1. Food price inflation, 2000–2004.

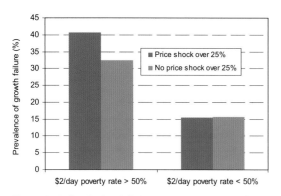

Figure 37.1. Food price inflation and undernutrition. *Sources:* World Bank, PovcalNet, n.d.; FAO 2008c.

when food prices rose more than 25% in any one year between 1998 and 2002. As the figure shows, countries with high poverty rates that experienced a price shock had higher growth failure rates than similarly poor countries that did not have a price shock. In the countries with less poverty, price shocks had no apparent impact on hunger. Map 37.1 shows that many countries with high poverty rates have had food price instability. Such countries include Zimbabwe, Myanmar, and Haiti, all of which experienced a doubling of food prices over the 2000–2004 period.

Since 2006, new demand for cereals to be used as fuel or feed rather than food has driven food prices to record-high levels around the world. Based on past experience shown in these maps and figures, recent price shocks are likely to translate into hunger for many poor people. While governments can intervene to address food price inflation, box 37.1 suggests that food price controls can be difficult to apply successfully. Instead, successful interventions are likely to be those that raise the real incomes and entitlements of the poor.

**Box 37.1. Food Price Dilemma**

Rising food prices can push vulnerable people into hunger. Faced with this prospect, well-intentioned governments sometimes impose controls to lower food prices, making essentials like bread affordable for the poor. In the first months of the year 2006, Mexico, Venezuela, and Zimbabwe each announced legal controls on the prices of certain food items. Unfortunately, food price controls often backfire, making food even less available to the poor.

Food price controls are problematic because government regulations to keep food prices low affect food suppliers and food consumers in inconsistent ways. Low official prices make food affordable to consumers, but at the same time these prices discourage food suppliers from producing and selling the food. The result is often empty shelves in the markets and in the homes of the poor. With the official price of bread in Zimbabwe set below the costs of baking it, the bakeries shut down and this basic food item became unavailable at any price (McGreal 2007). Similarly, when meat prices were held low in Venezuela, the most popular cuts of meat were soon unavailable in markets (Pearson 2007; Carroll 2007). By driving down supply, the use of price controls to make food affordable often results in greater scarcity and less access to food.

Controlling food prices poses a serious dilemma, since both high and low prices can worsen hunger outcomes. At too high a price food is unaffordable, but at too low a price, it is unavailable. This problem can be avoided by using policy interventions that affect food consumers and food producers in consistent ways. For example, programs like food stamps raise the purchasing power of the poor and enable vulnerable consumers to buy more food while also increasing revenues to food producers. Similarly, by removing a value-added tax on foodstuffs, the government of Venezuela recently made food more affordable to buyers without making it less profitable for suppliers. Interventions that create more employment and income for the poor will also enable more households to buy food without diminishing the returns to producing it.

# 38: Development Aid and Food Aid

Development aid can help end hunger in poor countries, if sufficient funds are donated and they are directed at improving the living standards of the poor. However, the rich countries of the world are not as generous as they could be, and the poor are rarely the main beneficiaries. Although the United States is the world's largest aid donor in absolute terms, in proportion to its wealth it is one of the least generous. In 2004, the ratio of development assistance to gross national income was 0.17%, far below the United Nations target of 0.7%. If the Millennium Development Goals are to be achieved by 2015, the wealthy countries of the world must commit more aid to programs that improve health, education, and infrastructure for long-term development.

Rich countries not only give insufficient aid, but the poorest countries are not at the top of the list of recipients. Aid monies typically flow to countries and regions based on the foreign policy interests of the donor. Just 10 countries received 61% of total US economic assistance in 2002. Of these, only three were low-income countries (Afghanistan, India, and Pakistan). The Middle Eastern/Southwest Asian regional focus of the aid reflects US geostrategic interests in the region. A similar foreign policy bias can be seen in the geographical distribution of development aid from Japan (Southeast Asia) and France and Great Britain (former colonies, overseas territories). When we compare the distribution of development assistance in map 38.1 with that of hunger vulnerability (map 8.1), there is no clear relationship between the two. This observation raises fundamental questions about the commitment of rich countries to achieving the Millennium Development Goal of halving hunger in the world.

If development aid is to improve food security, it must be directed toward the long-term development of countries where hunger vulnerability is highest. But

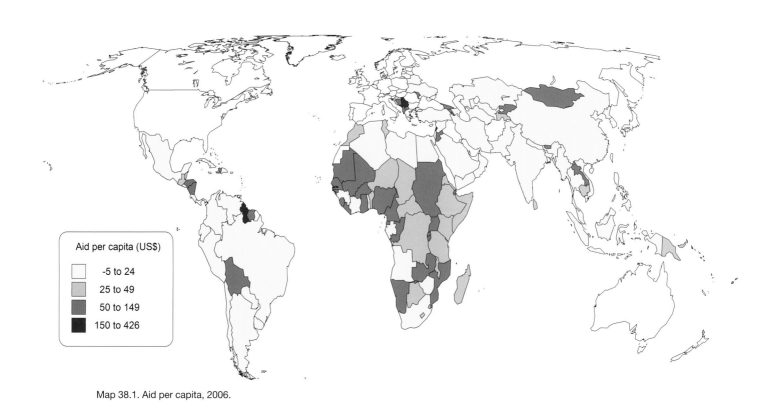

Aid per capita (US$)
- -5 to 24
- 25 to 49
- 50 to 149
- 150 to 426

Map 38.1. Aid per capita, 2006.

**Box 38.1. Food Aid Controversy in the United States**

The politics of US food aid made the news in the fall of 2005 when the head of the US Agency for International Development, which administers the US food aid program, argued that a quarter of the $1.2 billion food aid budget should be given in cash to food-insecure countries. There were two points to his argument. First, cash-based food aid would allow developing-country governments to purchase food from farmers in food surplus regions. This inflow of money would stimulate rural economies and investments in agriculture that are essential for long-term development. Second, buying food locally lowers transportation costs. For example, the United States sent 100,000 tons of grain worth $57 million to northern Uganda in 2003. Had the United States given the money directly to the Ugandan government, it could have bought food from local farmers in surplus-producing regions of the country. A Ugandan grain trader estimated that it cost the US food aid program $447 for each ton of corn shipped to Uganda. If the food had been purchased locally, it would have cost $180 per ton (Thurow and Kilman 2005). The savings from reduced shipping costs could have been used to purchase more food for Uganda's food and nutrition programs. Local purchases are also more timely. Delays of up to five months are common from the time an appeal for food assistance is made to its actual delivery by donors (Barrett and Maxwell 2005).

The food aid lobby fiercely resists cash-based food aid. NGOs like World Vision and Feed the Children, Inc., sell food aid in local markets to finance their aid programs. But food policy analysts have long criticized this practice because it depresses local prices and thus farmers' income. Shipping and handling companies also have a big stake in the food aid business. Food processing, packaging, and transportation accounts for 40% of total food aid costs (Murphy and McAfee 2005, 28). Agribusiness giants like Cargill and Archer Daniels Midland form a third group active in the food aid lobby. These grain traders and food processors receive above-market prices for the grain they sell to US food aid programs. All three groups lobbied the Senate and House agricultural committees in 2005 to stymie USAID's food aid reform initiative. They succeeded. Cash-based food aid was not included in either the House or Senate agricultural committee appropriation bills (Thurow and Kilman 2005). The food aid lobby made a minor concession in the 2008 Farm Bill. The legislation authorizes a four-year $60 million pilot program to study the effects of buying and distributing food in local and regional markets during food crises (Congressional Research Service 2008).

only a small portion of development assistance goes toward agricultural development, building schools and clinics, and employing teachers and doctors—the kind of investments that are necessary for transforming societies and economies in the long term (Sachs 2005). Development aid invests more in achieving geostrategic goals than in enabling people to move themselves out of poverty. When money does go to poor countries, it is usually for disaster relief and food aid. Though they may be valuable, these aid programs are not designed to promote long-term food security in the recipient country.

Even when it saves lives, development assistance also helps some donor-country interest groups who supply the goods and services funded by aid programs. The practice of sourcing goods and services in the donor country is known as "tied aid." For example, US food aid legislation requires that 75% of food aid come from US farms and that it be transported on US flag–carrying ships. As a result of this policy, the US food aid program benefits a small number of agribusiness firms, shipping companies, and NGOs (Murphy and McAfee 2005). This alliance's stake in the hunger-aid business became clear in 2005 when they successfully lobbied Congress to reject a USAID proposal to use 25% of its food aid budget to buy food in countries where food shortages and surpluses coexist (box 38.1).

Food aid analysts agree that buying food in the country needing food assistance benefits local farmers. Buying food grown in donor countries and shipping it overseas has the opposite effect. It can flood local markets, driving down prices and undermining local production. It is now considered best practice among donors to give cash instead of food because of long-

Figure 38.1. Elementary school students in the Korhogo region of northern Côte d'Ivoire participating in a school lunch program funded in part by the World Food Programme of the United Nations. The rice and oil is donated by the WFP, but students pay a small fee to cover the costs of transporting the food and to pay the cook.

term development benefits and because it is far less costly to buy food locally or in a neighboring country (Thurow and Kilman 2003). The United Nations Food and Agriculture Organization calls for terminating "tied" food aid and procuring more food aid in local and regional markets "to stimulate local production, strengthen local food systems and empower recipients in ways that traditional food aid cannot" (FAO 2007c). European Union countries are increasingly giving cash for local food purchases in their food aid programs. In 2003, cash-based food aid comprised 58% of the UK's and 81% of Germany's food aid budgets (Murphy and McAfee 2005, 12). This is an encouraging trend for those who hope that development aid can contribute to ending hunger. Opposition to conventional food aid is now arising from agencies that historically relied

heavily on it. In August 2007, the aid agency CARE announced that it would reject an offer of food aid worth 46 million dollars a year from the US government on the grounds that such aid actually did more harm than good.

Emergency food aid helps millions of people each year recover from natural and human-made disasters. School lunch programs and food supplements to pregnant women provide vital nutrients at the critical growth stages of women and children (figure 38.1). But as a solution to chronic hunger, food aid by itself can do little.

There are basically three types of food aid. *Emergency food aid* is destined for people affected by natural and human-caused disasters. It is distributed freely to disaster victims. *Project food aid,* such as food-for-work projects, aims to reduce poverty and prevent disasters. Donated food is either distributed freely to specific groups or sold in the open market for revenue used to fund aid projects. Both emergency and project food aid are given as grants to recipient countries and often are administered by NGOs. *Program food aid* seeks to assist governments experiencing financial difficulties by providing food through grants or loans. The food is then sold in open markets and the proceeds used to reduce budget deficits or balance-of-payments problems. In the three-year period 2002–4, emergency food aid accounted for 55% of all food aid, up from 39% in 1996–1998.

Food surplus countries either contribute food to international organizations like the UN's World Food Programme or distribute it directly to countries through national food aid programs. About two-thirds of food aid is transferred directly from donor countries and organizations to recipient countries. The remainder is purchased within the recipient country or region with cash provided by donors.

Historically, food aid was designed to assist farmers in the developed world in disposing of agricultural surpluses (figure 38.2) This transfer of "cheap food" from surplus-producing countries to food-deficit countries had a secondary goal of creating long-term trade relationships between donor and recipient countries. "From-aid-to-trade" is a slogan used by aid administrators to persuade government legislators to fund aid programs because they benefit donors as much as recipients (see box 38.1). A third of the global food aid budget, some $600 million, is spent in donor countries (FAO 2007c).

Since a large share of food aid is given in emergency situations, it does little to address chronic hunger. However, when combined with nonfood assistance and public health programs that do tackle poverty, improve nutrition, and reduce disease, food aid can play an important role in reducing hunger vulnerability.

But how do we know if food aid actually reaches those in need? The "targeting" of food aid is of major concern to food aid program administrators. Getting food aid to afflicted areas and households, and making sure that it does not go to people who are not in need, are basic issues. We do not have data at the subnational scale to know who is included in and excluded

Figure 38.2. A bumper harvest of corn waiting to be moved to a silo in Champaign County, Illinois, October 2005.

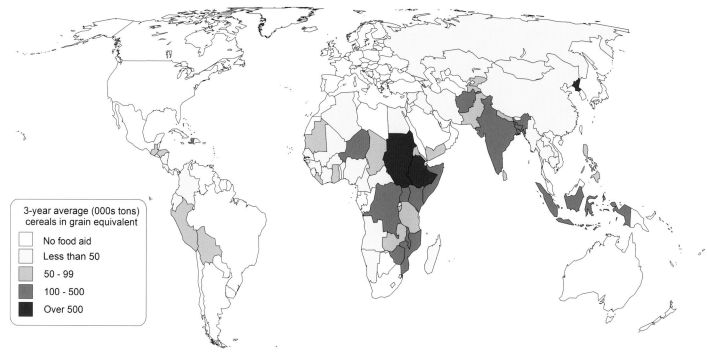

Map 38.2. Food aid deliveries, 2005–7.

**Table 38.1. Food aid deliveries by region, 2005–7 average**

| Region | Percentage of global food aid |
|---|---|
| Sub-Saharan Africa | 55.5 |
| Asia | 26.8 |
| Latin America & the Caribbean | 7.7 |
| Middle East & North Africa | 5.1 |
| Eastern Europe & Commonwealth of Independent States | 5.1 |

Source: WFP 2008

**Table 38.2. The top 10 recipients of food aid, 2005–7 average**

| Country | Cereal grain equivalents (000 tons) |
|---|---|
| Ethiopia | 825 |
| North Korea | 708 |
| Sudan | 666 |
| Uganda | 267 |
| Kenya | 260 |
| Bangladesh | 248 |
| Afghanistan | 197 |
| Malawi | 158 |
| Palestinian Territory | 122 |
| Zimbabwe | 121 |

Source: WFP 2008

from food aid programs. Map 38.2 shows food aid deliveries for 2005–7 by country. In terms of total food aid (cereal grain equivalents) delivered, Sub-Saharan Africa received 55.5% and Asia 26.8% of all food aid (table 38.1).

On the basis of this regional distribution, food aid appears to be reaching those parts of the world experiencing the greatest hunger vulnerability (map 8.1). Food aid within these regions is often concentrated in a small number of countries (map 38.3). Just three Sub-Saharan countries (Ethiopia, Sudan, and Uganda) received 45% of all food aid to that region in 2005–7. North Korea alone accounted for 38% of Asia's total. On average, the top 10 recipients of food

aid received 53% of all food distributed in 2005–7 (table 38.2).

To determine whether the amount of food aid received by a country is at all related to the number of its poor citizens, we divided the amount of food aid by the number of people living in extreme poverty. If food aid deliveries were based on need, we would expect to find similar amounts of food aid per poor person across the world. Map 38.4 shows the number of kilograms of food aid available per person living in extreme poverty for the three-year period 2005–7. We do not know if

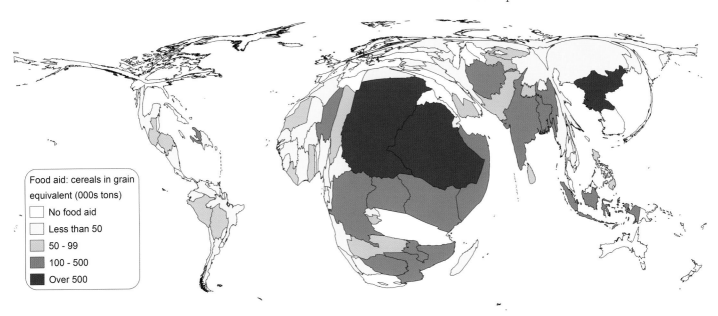

Map 38.3. Food aid recipients, 2002–4 average.

Food aid: cereals in grain equivalent (000s tons)

- No food aid
- Less than 50
- 50 - 99
- 100 - 500
- Over 500

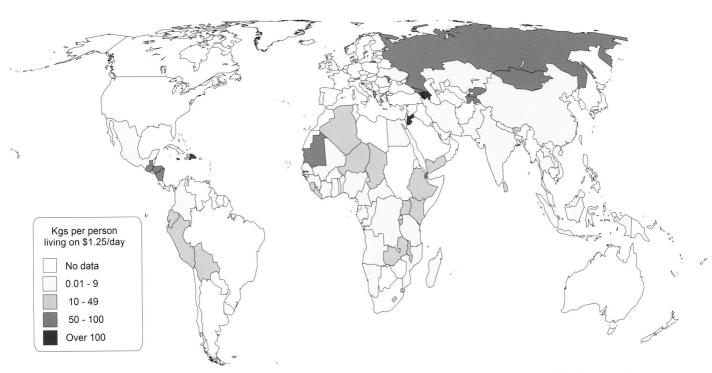

Map 38.4. Food aid per person living in extreme poverty, 2004–7, three-year average. *Note:* Poverty data missing for Iraq, North Korea, Lebanon, Myanmar, Somalia, Sudan, Syria, and Zimbabwe.

Kgs per person living on $1.25/day

- No data
- 0.01 - 9
- 10 - 49
- 50 - 100
- Over 100

food is actually getting to needy people within these countries. What the map shows, however, is wide variation in the number of kilograms received per needy person. In India, less than 1 kilogram of food aid was delivered per person in extreme poverty, while Azerbaijan received over 19,500 kilograms per poor person.

Ethiopia received 28 kilograms per person in extreme poverty, while in Jordan the amount surpassed 1900 kilograms. This uneven geography of food aid suggests that factors other than need play a role in the global distribution of food aid.

# 39: War

Previous entries in this section examine how certain conditions in households or national economies make people more vulnerable to hunger. These final maps consider sudden disastrous events that often move functional but poor households into destitution and hunger.

Armed conflicts produce hunger when they disrupt food systems, impoverish households, hamper food relief, and divert scarce public resources from long-term development to military expenditures. Contemporary wars disproportionately affect citizens of poor countries. Most of the three million deaths related to armed conflicts between 1990 and 2004 took place in low-income countries. In 2005 another 25 million people were internally displaced due to conflicts and human rights abuses (UNDP 2005, 151). More than two-thirds of the 32 countries in the UN's Low Human Development category have suffered an armed conflict since 1990. There is truth to the dictum that "Poverty fosters war, and war impoverishes" (*Economist* 2003, cited in O'Loughlin 2005, 88).

War turns food shortages into famines. This was evident between 1983 and 1985 when famine struck many countries in Western, Northeastern, and Southern Africa, where hunger vulnerability was worsened by severe drought and war. By 1987 food security had greatly improved in Western Africa, but civil wars in the Northeast (Ethiopia, Sudan) and South (Angola, Mozambique) continued to produce widespread hunger.

In both regions, politically motivated violence devastated the countryside, where crops were destroyed, roads and bridges were damaged, and schools and clinics were frequent targets of fighting. Even food aid was used as a weapon against people living in rebel-controlled areas (de Waal 1997). In many regions it was too dangerous for relief workers to count the hungry and deliver food supplies. The use of food as a political weapon was repeated in Zimbabwe in 2005 and in Ethiopia in 2007, when emergency food distribu-

tions were prohibited in areas that opposed the central government.

The more governments spend on their military, the less they have available for social services. In the late 1980s governments in Africa commonly devoted 30%–40% of their national budgets toward warfare rather than citizen welfare. (The US share of defense spending in the federal budget was 27% during the same period.) In 2004, the world average share of military expenditures amounted to 11% of government budgets. Map 39.1 shows the share of military spending in central government budgets for the most recent year for which data are available. Many countries with critical hunger problems continue to spend well above this world average.

Wars also generate massive refugee populations whose capacity to produce or buy food is greatly undermined by their limited access to jobs and food-producing resources. Dislocation and heightened vulnerability to disease result in higher levels of infant and child deaths among war refugees. This was the case in the in the war-torn eastern province of the Democratic Republic of the Congo in the period 1998–2004. There the infant mortality rates were double the rate for Sub-Saharan Africa and 70% higher than the national average (UNDP 2005, 156). Maps 39.2 and 39.3 reveal the large numbers of people displaced by war, political instability, and other human-caused disasters. They include refugees, returnees, asylum seekers, internally displaced persons, and groups of war victims, all of whom are of concern to the UN High Commission on Refugees. Regional wars in West and Central Africa and Southwest Asia are major sources of these displaced populations. These are also areas of extreme hunger vulnerability.

Map 39.4 shows the distribution of armed conflicts in the world between 2001 and 2005. A major armed conflict is defined as one in which more than 100 people died within a year and more than 1000 were killed over

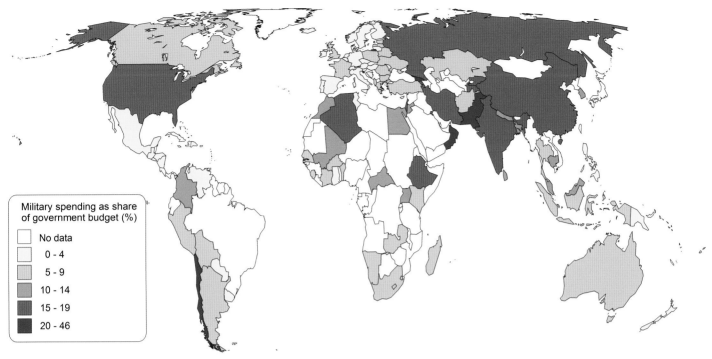

Map 39.1. Military spending as share of government budget, 2000–2006.

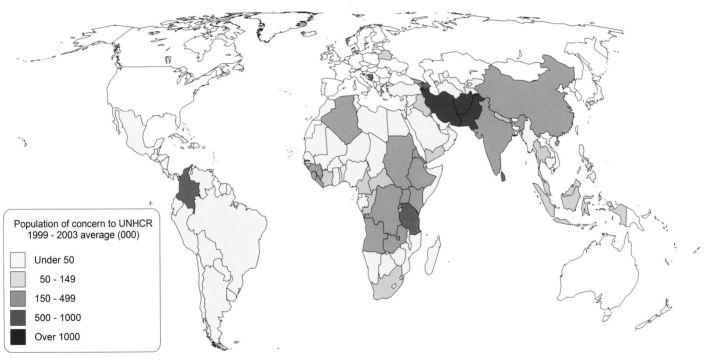

Map 39.2. People displaced by human-made disasters, 1999–2003, five-year average.

the life of the conflict (Marshall and Gurr 2005, 77). The magnitude scale refers to the destructive impact, both human and physical, of the conflict. It ranges from a low-impact indicator of 1 to a total destruction indicator of 10.

Whether war precipitates a hunger crisis depends on a country's food security status. A comparison of map 39.1 with the geography of hunger vulnerability (map 8.1 reveals that war is more likely to result in a food crisis in countries where hunger vulnerability is already high. For example, both Russia and Sudan experienced armed conflicts of the same magnitude in 2005. However, only Sudan experienced a food crisis. Its much higher level of hunger vulnerability (HVI = 42) in comparison to Russia (HVI = 8) forced millions

of poor people in the conflict-ridden Darfur region to seek food aid from the World Food Programme in 2005. The World Health Organization estimated that 70,000 people died in Darfur from malnutrition and disease over a seven-month period in 2004 (Lacey 2005). The

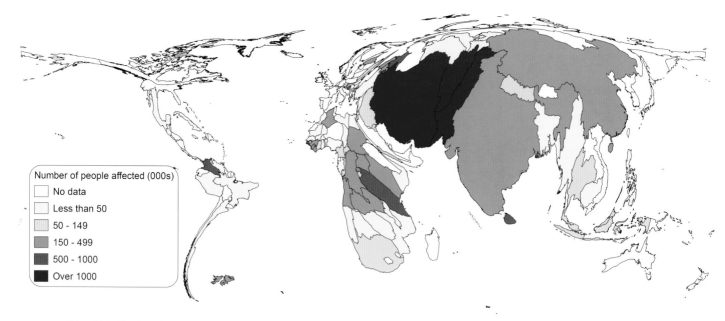

Map 39.3. Number of people affected by human-made disasters, 1999–2003, five-year average.

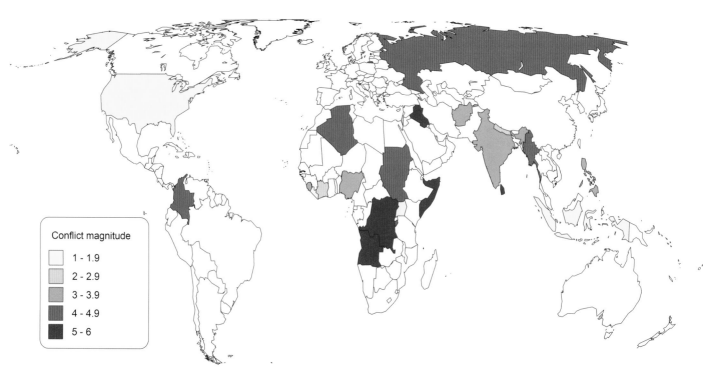

Map 39.4. Major armed conflicts, 2001–5. *Note:* The magnitude scale refers to the destructive (human and physical) impact of the conflict. It ranges from a low-impact indicator of 1 to a total destruction indicator of 10.

Figure 39.1. United Nations peacekeeping troops patrol the northern city of Korhogo, Côte d'Ivoire, in March 2008. From 2002 to 2009, when a rebellion divided the country, UN forces provided food, water, and public security to civilians caught in the conflict. The combined humanitarian and military assistance helped to reduce hunger vulnerability among the population.

war between the Russian government and Chechen separatists left hundreds dead, but hunger did not claim lives as in Sudan.

In 2004 there were 35 countries that received emergency food aid across the world. Almost half of these "hunger hotspots" were involved in a current or past conflict (FAO 2004a, 16). It is abundantly clear that resolving civil conflicts is a necessary condition for food security in these areas (figure 39.1).

# 40: Natural Disasters

Some disasters, like war and political instability, are clearly human-made. Other disasters have more natural origins, such as droughts, volcanic eruptions, earthquakes, floods, and tidal waves (tsunami). Each year people are affected by these natural events that destroy homes, crops, and livelihoods. Whether they turn into disasters depends on the preparedness and vulnerability of the affected populations. The rapidity of emergency relief efforts is also critical in determining whether thousands die or suffer in the aftermath of these events. The rapid and generous international aid response to the Indian Ocean tsunami of December 26, 2004, that killed nearly 200,000 people and left two million homeless prevented a major disease outbreak and famine among the displaced populations across the region. The relatively slow response to the drought and locust invasion that reduced crop yields and cattle fodder in Niger in 2005 heightened vulnerability for

1.2 million farmers and livestock producers—about a third of the country's rural producers (Wines 2005). Poor farmers and herders, particularly their children, were most affected by the drought and pestilence. In "normal" years, a quarter of Niger's children die before reaching the age of five. When natural calamities strike and social safety nets are weak, children die in even larger numbers.

A natural disaster can impoverish people instantly, but political and economic factors determine whether the victims recover. Unless emergency food, shelter, and health care are made available, the poor find it difficult to cope. Market shocks can follow and compound the impacts of natural disasters, wreaking havoc on rural producers whose livelihoods depend on buying and selling goods. When many desperate people sell livestock and other assets during a drought, the prices of these goods decline just as food prices rise due to local

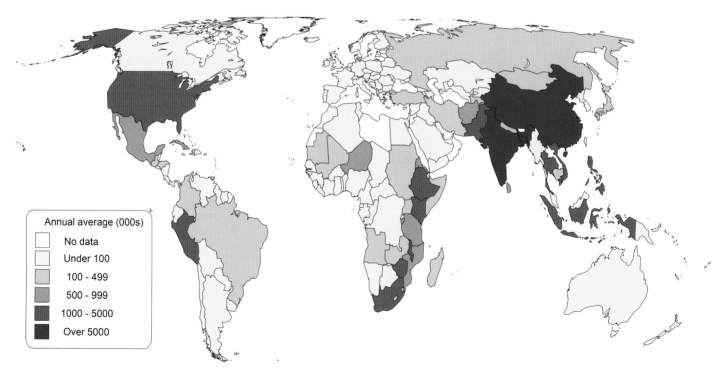

Annual average (000s)

- No data
- Under 100
- 100 - 499
- 500 - 999
- 1000 - 5000
- Over 5000

Map 40.1. Number of people affected by natural disasters, 2003–6.

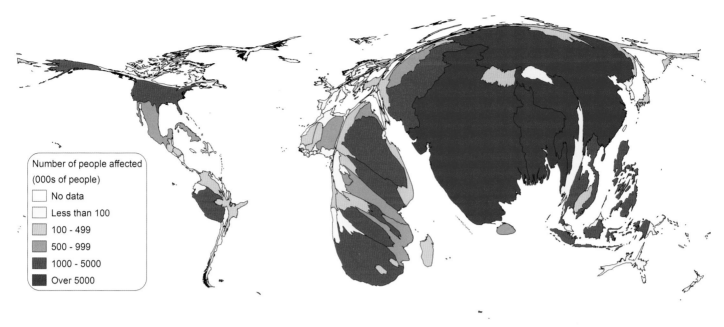

Map 40.2. Number of people affected by natural disasters, four-year annual average, 2003–6.

scarcities. Government intervention is critical to buffer such households from the double burden of natural disasters and economic turmoil.

Maps 40.1 and 40.2 show the number of people affected by natural disasters over the four-year period 2003–6. "Affected people" refers to individuals requiring food, water, shelter, sanitation, and medical assistance following a natural disaster. Sometimes infectious diseases spread into areas where they were formerly absent due to a natural disaster. People who become ill from these disaster-related diseases are also counted.

Most people affected by natural disasters reside in East and South Asia and Sub-Saharan Africa. China alone accounted for 62.8% of the world total, followed by India (11.3%), Bangladesh (5.5%), South Africa (2.1%), Pakistan (2%), Ethiopia (1.9%), and the Philippines (1.7%). Worldwide, 90% of the people needing assistance following a natural disaster resided in just 10 countries. The percentage of the national population in need of help reached high levels in China, Bangladesh, South Africa, and Zimbabwe (table 40.1). Flooding wreaked the most havoc. Fifty-five percent of the world's population that required emergency assistance in 2003–6 experienced floods. Windstorms and droughts accounted for 21.0% and 20.6% of the population affected by disasters. Natural disasters dispropor-

**Table 40.1. Number of people affected by natural disasters, 2003–6**

| | Annual average | % World total | % National population |
|---|---|---|---|
| China | 111,296,282 | 62.8 | 8.6 |
| India | 19,946,385 | 11.3 | 1.8 |
| Bangladesh | 9,714,079 | 5.5 | 7.0 |
| South Africa | 3,755,873 | 2.1 | 8.3 |
| Pakistan | 3,470,237 | 2.0 | 2.3 |
| Ethiopia | 3,355,094 | 1.9 | 4.8 |
| Philippines | 2,997,293 | 1.7 | 3.7 |
| Kenya | 1,673,434 | 0.9 | 5.0 |
| United States | 1,577,103 | 0.9 | 0.5 |
| Zimbabwe | 1,504,983 | 0.8 | 11.6 |

Source: EM-DAT.

tionately hurt the poor, who have fewer resources to buffer them from the calamity. Even in rich countries like the United States, the poor and ethnic minorities suffer most from natural disasters. When Hurricane Katrina destroyed much of New Orleans on August 29, 2005, low-income African American households had fewer options than middle-income white households to escape and recover from the hurricane and subsequent flooding of the city (Logan 2007).

Both vulnerability to and recovery from natural disasters depend on the capacity of households to withstand nature's shocks, and the assistance they receive from others to buffer them from its worst effects. That a prosperous country like the United States could fail to respond adequately to a natural disaster as occurred in New Orleans suggests how difficult effective responses must be in poor countries whose resources are more limited and households more vulnerable to hunger even in normal circumstances. One likely consequence of global climate change (map 18.1) is an increased frequency of floods, hurricanes, and other dramatic climatic events. These shocks are likely to strain the abilities of poor countries to support their populations in times of crisis and exacerbate hunger unless international responses are exceptionally effective.

# 41: Conclusion

This atlas is an inquiry into the status of hunger in the world. It documents evidence of hunger and examines relationships between the geography of hunger vulnerability and its possible sources. Each map, in this sense, is a question that explores the relationship between world hunger and its many causes. This exploration is necessarily limited by the availability of data and, by extension, the limits of empiricism. We cannot reduce the processes and patterns of hunger to discrete measurable factors, which when added up in check-list fashion, offer an explanation.

Hunger is commonly thought to result from overpopulation, human and natural disasters, geographical factors, ignorance of good nutrition, or some combination of these factors (Sachs 2003; World Bank 2006). In our conceptual framework, hunger is rooted in social vulnerability. People who are chronically hungry or at risk of becoming hungry are typically the poor, who lack access to food-producing resources and/or incomes to obtain the food they need when they need it. Our hunger vulnerability framework links hunger to poverty through the notion of entitlements, or a person's ability to cover his or her basic needs through production, employment, and access to public and private services. Vulnerability hinges on the political, economic, and social processes that undermine or support a population's ability to escape chronic hunger or respond to extreme events (Watts 1987; Ribot 1996). Poor people can avoid hunger and cope with disasters and market shocks if they possess sufficient assets (grain reserves, livestock, savings) and can make claims to the resources held by others such as family, community, government, and international organizations. But when these assets, social networks, and claims are insufficient to sustain the poor through particularly hard times, hunger can result. We can trace hunger vulnerability to specific factors that intensify poverty and reduce people's en-

titlements. This approach points to interventions and institutions that will minimize the poor's exposure to hunger (Sen 1981; Ribot 1996).

The Brazilian government's Zero Hunger Project attacks hunger vulnerability through a multiprong approach that combines emergency assistance and increased access to affordable food with asset redistribution via agrarian reform, rural infrastructure investments, and greater support for family farms (FAO 2004a, 22). As in Uruguay (see maps 9.1–9.3), Mexico's Education, Health, and Nutrition Program (PROGRESA) supports rural families living in extreme poverty through school lunch programs. The program provides meals in schools and monetary assistance to mothers as long as they attend health clinic programs and keep their children in school (Sanchez et al. 2005). If school lunch programs obtained their food from local agriculture, then low-income farmers could improve their food security as well. Synergies like these that strengthen rural economies, improve school attendance, and provide direct access to food help to reduce hunger vulnerability where it is greatest. In India, where half of the children under five suffer undernutrition, the Integrated Child Development Program seeks to improve the nutritional status of children under two years of age. It combines antenatal care with provision of micronutrient supplements to pregnant women, and stresses exclusive breastfeeding for the first six months of a child's life. Children under the age of three also receive supplements of Vitamin A, iron, and folic acid that are critical for child growth (Bread for the World 2006, 92).

Studies from dozens of countries in Asia, Latin America, and Sub-Saharan Africa show significant reductions in child malnutrition with improvements in the status of women. When women's decision-making power increases within households and communities and they control more income, more money is spent on child health, education, and nutrition. According to many studies, women's empowerment, often through education, is twice as likely to improve children's nutritional levels as increased food availability within a country (Kennedy and Peters 1992; Smith et al. 2003; Bread for the World 2006, 90–91).

Despite the increased awareness of and commitment to ending hunger in the developing world, the trends are not encouraging. Little progress has been made in achieving the 1996 World Food Summit (WFS) goal of cutting in half the *number* of hungry people in developing countries by the year 2015. Between the baseline period of 1990–1992 and 2002–4, the number increased by seven million, rising from 823 to 830 million people. This worsening trend is even more worrisome since the number of undernourished people in the developing world had dropped to 797 million in the period 1995–97 (FAO 2006b).

Although the number of hungry people is rising, there has been progress toward the related Millennium Development Goal of halving the *proportion* of the world's population that is undernourished, from 20% to 10%, by 2015. The data show a decline from 20% to 17% between 1990–92 and 2002–4. The percentage of people suffering hunger declined while the number of hungry people increased, because the total population is growing faster than the hungry population. As the FAO notes, meeting the World Food Summit objective of halving the number of the hungry requires the proportion of hungry in the world to decline by more than half (FAO 2006b).

Tables 41.1 and 41.2 and maps 41.1 and 41.2 show how progress can be mapped differently depending on which performance measure is used to track hunger trends. Table 41.1 indicates that between 1990–92 and 2002–4 the number of undernourished people increased by 21% in East Africa and 7% in Southern Africa. The number of hungry increased by 3% in South Asia. If we consider the percentage of the population in hunger rather than their number, table 41.2 indicates that hunger declined by 11% and 19% in East and Southern Africa, and by 19% in South Asia. When we map these different measures (maps 41.1–41.2), it appears that we are looking at two different worlds. Which one do we use to chart the state of global hunger?

Different indicators will give different results, as illustrated in maps 41.1 and 41.2. This example illustrates how maps can be used to make an argument, to per-

**Table 41.1. Change in number of undernourished people in the developing world by region, 1990–92 and 2002–4**

| Region | 1990–92 (millions) | 2002–4 (millions) | % Change |
|---|---|---|---|
| **Sub-Saharan Africa** | **169.0** | **213.4** | **26.0** |
| Central Africa | 22.7 | 49.1 | 116.0 |
| East Africa | 75.1 | 91.2 | 21.0 |
| Southern Africa | 34.1 | 36.5 | 7.0 |
| West Africa | 37.2 | 36.6 | –1.6 |
| **Latin America and Caribbean** | **59.4** | **52.1** | **–12.3** |
| North America (Mexico) | 4.6 | 5.3 | 15.0 |
| Central America | 5.0 | 7.5 | 50.0 |
| Caribbean | 7.7 | 6.8 | –12.0 |
| South America | 42.0 | 32.5 | –22.6 |
| **Asia and Pacific** | **569.7** | **527.0** | **–7.5** |
| East Asia | 198.7 | 163.0 | –18.0 |
| Southeast Asia | 80.0 | 64.0 | –20.0 |
| South Asia | 290.4 | 300.0 | 3.0 |
| **Near East and North Africa** | **25.0** | **37.3** | **49.0** |
| Near East | 19.6 | 31.4 | 60.0 |
| North Africa | 5.4 | 5.9 | 9.3 |
| DEVELOPING WORLD | **823.1** | **829.8** | **0.9** |

Source: FAO 2008a.

**Table 41.2. Change in the prevalence of undernourishment in the developing world by region, 1990–92 and 2002–4**

| Region | 1990–92 (%) | 2002–4 (%) | % Change |
|---|---|---|---|
| **Sub-Saharan Africa** | **35** | **33** | **–5.7** |
| Central Africa | 36 | 57 | 58.3 |
| East Africa | 45 | 40 | –11.1 |
| Southern Africa | 48 | 39 | –18.8 |
| West Africa | 21 | 15 | –28.6 |
| **Latin America/Caribbean** | **13** | **10** | **–23.1** |
| North America(Mexico) | 5 | 5 | 0 |
| Central America | 17 | 19 | 11.8 |
| Caribbean | 27 | 21 | –22.2 |
| South America | 14 | 9 | –35.7 |
| **Asia/Pacific** | **20** | **16** | **–20.0** |
| East Asia | 16 | 12 | –25.0 |
| Southeast Asia | 18 | 12 | –33.3 |
| South Asia | 26 | 21 | –19.2 |
| **Near East and North Africa** | **8** | **9** | **12.5** |
| Near East | 10 | 12 | 20.0 |
| North Africa | 4 | 4 | 0 |
| DEVELOPING WORLD | **20** | **17** | **–15.0** |

Source: FAO 2008a.

suade the reader that the world is like this, and not like that (Harley 1989; Wood 1992). One of the objectives of this atlas is to illustrate the difficulty of locating and explaining hunger in the world, and to offer the Hun-

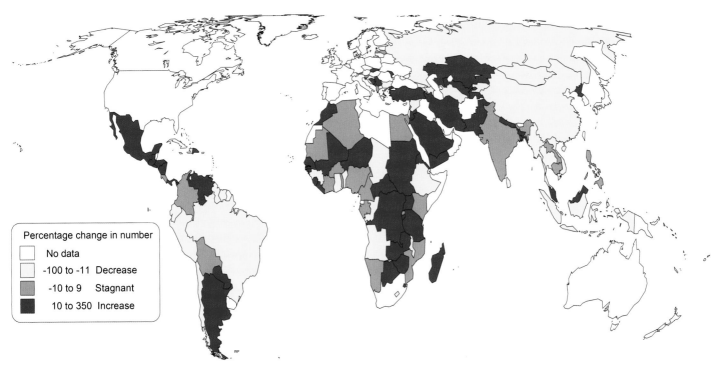

Map 41.1. Change in number of undernourished people, 1990–92 to 2002–4.

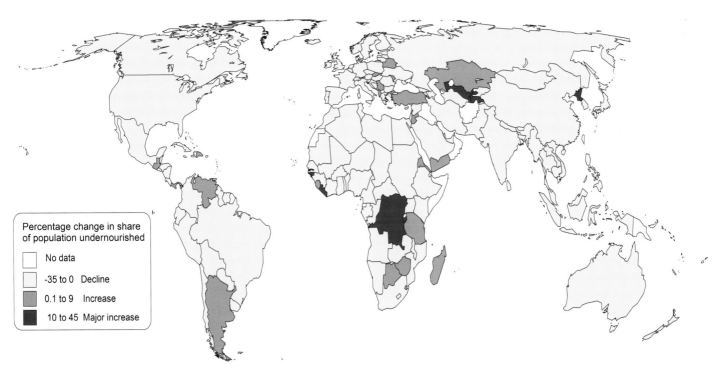

Map 41.2. Change in proportion of population undernourished, 1990–92 to 2002–4.

ger Vulnerability Index as an alternative measure. The HVI is one of many indicators. We hope that this atlas will spur its readers to reflect on additional measures that will advance our understanding. We also hope that they will become critical readers of maps.

## LOCATING AND EXPLAINING HUNGER: WHAT HAVE THE MAPS SHOWN?

The maps in the first section of this atlas demonstrate where people suffer hunger and vulnerability to hunger. Based on what we believe to be the causes of hunger, the later maps examine the factors that may be responsible for hunger vulnerability. The first maps tried to show where hunger exists, while the later maps pointed to what must be changed to end hunger vulnerability.

The maps explaining the roots of hunger were chosen based on the idea that hunger vulnerability is a result of poverty, and poverty is rooted in the interplay of national resources, available technology, and power relations expressed through market and political institutions. Many variables have been mapped to capture these underlying sources of hunger, but the effort to identify which factors are especially responsible as causes can be daunting. In some cases there is very little correlation between the distribution of some factor often thought to cause hunger, like population growth, and hunger vulnerability. In other cases, there is a strong correlation between distribution of hunger and the distribution of some condition, but we cannot be certain which condition causes the other or if the relationship is merely a coincidence. While we cannot confidently discern causes from coincidences in these maps, we can conclude that factors whose distribution around the world is only weakly correlated to hunger are not powerful independent causes of the problem. They may be significant causes when combined with other factors.

To help bring some clarity to the maps, table 41.3 classifies which factors are strongly correlated with hunger vulnerability, which ones are strongly correlated among some specific groups of countries, and which seem weakly associated with the problem. Excluded from

**Table 41.3. Correlation between hunger vulnerability and different variables**

*Strong correlation*
   Health care
   Literacy
   Technology
   Gender equality
*Strong correlation, under some conditions*
   National income per capita (at low income levels)
   Income distribution (at lower-middle income levels)
   Land per capita (among low-income, agricultural countries)
*Weak relationship*
   Population growth
   Health of the environment
   Changes in resource base
   Road density
   Political freedoms

this table are the many factors identified as exacerbating conditions. Catastrophic events like war and chronic conditions of poor health all contribute to hunger among the vulnerable. Here we concentrate on the basic factors that may create that vulnerability. Also excluded are the components of the Hunger Vulnerability Index (food availability, poverty, and growth failure).

Among those factors that are consistently correlated with hunger vulnerability are two measures of human resources (literacy rates and health care expenditures), the measures of technology, and gender equality. In this group we see indication of the roles of resources, technology, and social relations.

Three factors appear to correspond to increased hunger among certain types of countries. When average incomes are low, national income per capita is closely related to hunger. Lifting total income in the poorest countries corresponds to reduced vulnerability. For a second group of countries that have slightly higher but still quite low average-income levels, the distribution of income is closely associated with hunger. Among these low and lower-middle-income countries, greater inequality corresponds to greater hunger. Finally, falling land availability corresponds to greater hunger and poverty in those countries where a large share of the population depends on land for their income.

Most of the measures of environmental stress were weakly correlated with hunger. Our measures of physical infrastructure and of aggregate political freedom were also poorly correlated with vulnerability. In some cases, it may be that concepts like political freedom or environmental stress were so poorly measured that an existing relationship is simply not observed. However, the information at hand suggests little role for population as a cause of hunger, and a small role for the physical environment, except where the population depends on land for income. In contrast, we see a potentially large role for human resources in terms of health and education, technology availability and creation, and women's rights.

In summary, this atlas supports the general argument that hunger is closely tied to poverty and social vulnerability. Vulnerability is reduced when people are empowered to demand from the political economy the right to sufficient resources and entitlements to buffer them against hunger.

# Appendix 1: Map Data Sources

Map 1.1 (Safe drinking water): UNICEF 2008, http://www.unicef .org/wes/mdgreport/globalEstimate.php; World Bank, World Development Indicators Online, http://go.worldbank.org/ 6HAYAHG8H0.

Map 1.2 ($1.25/day poverty): World Bank, World Development Indicators Online, http://go.worldbank.org/6HAYAHG8H0.

Map 2.1 (Food supply 2300 calories per day): FAO 2008b.table D1, 189–92, http://www.fao.org/statistics/yearbook/vol_1_1/ pdf/d01.pdf.

Map 2.2 (Change in food availability): FAO 2008b, http://www .fao.org/statistics/yearbook/vol_1_1/pdf/d01.pdf.

Map 2.3 (Food supply 2100 calories per day): FAO 2008b, table D1, 189–92, http://www.fao.org/statistics/yearbook/vol_1_1/ pdf/d01.pdf.

Map 3.1 (Prevalence of undernourishment): FAO 2006, Food Security Statistics, http://www.fao.org/faostat/foodsecurity/ index_en.htm (accessed September 1, 2008).

Map 3.2 (Change in undernourishment): FAO 2006, Food Security Statistics, http://www.fao.org/faostat/foodsecurity/index _en.htm (accessed September 1, 2008).

Map 3.3 (Number undernourished): FAO 2006, Food Security Statistics, http://www.fao.org/faostat/foodsecurity/index _en.htm (accessed September 1, 2008).

Map 3.4 (Change in number undernourished): FAO 2006, Food Security Statistics, http://www.fao.org/faostat/foodsecurity/ index_en.htm (accessed September 1, 2008).

Map 4.1 (Vitamin A deficiency): Center for Human Nutrition 2008, http://www.jhsph.edu/chn/images/GlobalVADtables .pdf.

Map 4.2 (Iodine): WHO 2004b, http://whqlibdoc.who.int/ publications/2004/9241592001.pdf, table A2.6.

Map 4.3 (Anemia): WHO 2008g, http://whqlibdoc.who.int/ publications/2008/9789241596657_eng.pdf.

Map 5.1 (Child obesity): WHO 2008c, http://www.who.int/ nutgrowthdb/database/countries/en/index.html.

Map 5.2 (Adult obesity): WHO Global Database on Body Mass Index 2008, http://www.who.int/bmi/index.jsp.

Map 6.1 (Growth failure): WHO 2008c, http://www.who.int/ nutgrowthdb/database/countries/en/index.html.

Map 6.2 (Change in growth failure rate): United Nations System, Standing Committee on Nutrition 2004, 71–72, http://www .unsystem.org/scn/publications/AnnualMeeting/SCN31/ SCN5Report.pdf.

Map 6.3 (Change in growth failure numbers): United Nations System, Standing Committee on Nutrition 2004, 71–72, http://www.unsystem.org/scn/publications/Annual Meeting/SCN31/SCN5Report.pdf.

Map 7.1 ($2/day poverty, percentage): World Bank, PovcalNet, 2008, http://iresearch.worldbank.org/PovcalNet/povcalSvy .html.

Map 7.2 ($2/day poverty, numbers): World Bank, PovcalNet, 2008, http://iresearch.worldbank.org/PovcalNet/povcalSvy.html.

Map 8.1 (HVI): WHO 2004c, http://www.who.int/nutgrowthdb/ database/countries/en/index.html; FAO 2008b, http://www .fao.org/statistics/yearbook/vol_1_1/pdf/d01.pdf; World Bank, PovcalNet, 2008, http://iresearch.worldbank.org/ PovcalNet/povcalSvy.html.

Map 9.1 (Growth failure by department, Uruguay): ANEP 2003, 64–67.

Map 9.2 (Rural-urban differences): ANEP 2003, 64–67.

Map 9.3 (Child growth failure, Montevideo): ANEP 2003, 70–71.

Map 10.1 (Food insecurity, USA): Nord et al. 2007, appendix D, table D-1, http://www.ers.usda.gov/Publications/ERR49/ ERR49appD.pdf.

Map 10.2 (Changes in hunger prevalence, USA): Nord et al. 2007, 55.

Map 10.3 (Poverty per capita, USA): USDA, Food Stamp Program Map Machine, http://www.ers.usda.gov/Data/FoodStamps/ (accessed November 11, 2008).

Map 10.4 (Percentage of poor in food stamp program, USA): USDA, Food Stamp Program Map Machine, http://www.ers .usda.gov/Data/FoodStamps/ (accessed November 11, 2008).

Map 11.1 (Severe anemia in India): IIPS and Macro International 2007, http://www.measuredhs.com/pubs/pdf/ FRIND3/10Chapter10.pdf, 290.

Map 11.2 (Child growth failure, India): IIPS and Macro International 2007, http://www.measuredhs.com/pubs/pdf/ FRIND3/10Chapter10.pdf, 273.

Map 11.3 (Adult obesity, Mexico): Barquera et al. 2006, http://www.fao.org/docrep/009/a0442e/a0442e0k .htm#TopOfPage, table 15.

Map 12.1 (Population): World Bank, World Development Indicators Online, 2008, http://go.worldbank.org/6HAYAHG8H0.

Map 13.1 (Arable land): World Bank, World Development Indicators Online, 2008, http://go.worldbank.org/6HAYAHG8H0.

Map 13.2 (Change in arable land): World Bank, World Development Indicators Online, 2005, http://go.worldbank.org/ 6HAYAHG8H0.

Map 14.1 (Environmental system): Esty et al. 2005, http://www .yale.edu/esi/d_comindtables.pdf, 352.

Map 15.1 (Literacy): World Bank, World Development Indicators Online, 2008, http://go.worldbank.org/6HAYAHG8H0; CIA 2008, The World Factbook, https://www.cia.gov/library/ publications/the-world-factbook/docs/profileguide.html;

UNESCO Institute for Statistics, Data Centre 2008, http://stats.uis.unesco.org/unesco/TableViewer/tableView.aspx?ReportId=210.

Map 16.1 (Road density): World Bank, World Development Indicators Online, 2008, http://go.worldbank.org/6HAYAHG8H0.

Map 17.1 (Resource base): World Bank, World Development Indicators Online, 2008, http://go.worldbank.org/6HAYAHG8H0.

Map 18.1 (Climate change and agriculture): Tubiello and Fischer 2007, 1046, table 4(b).

Map 19.1 (College enrollment): World Bank, World Development Indicators Online, 2008, http://go.worldbank.org/6HAYAHG8H0.

Map 19.2 (Research): World Bank, World Development Indicators Online, 2008, http://go.worldbank.org/6HAYAHG8H0.

Map 19.3 (Investment in R&D): World Bank, World Development Indicators Online, 2000–2006, http://go.worldbank.org/6HAYAHG8H0.

Map 19.4 (Digital divide): World Bank, World Development Indicators Online, 2008, http://go.worldbank.org/6HAYAHG8H0.

Map 20.1 (Fertilizer): World Bank, World Development Indicators Online, 2008, http://go.worldbank.org/6HAYAHG8H0.

Map 21.1 (Colonialism): Schrader and Gallouédec 1914.

Map 22.1 (Debt): World Bank, World Development Indicators Online, 2008, http://go.worldbank.org/6HAYAHG8H0.

Map 23.1 (Political freedoms): Freedom House 2008, http://www.freedomhouse.org/template.cfm?page=414.

Map 24.1 (Poorest 10%): World Bank, World Development Indicators Online, 2008, http://go.worldbank.org/6HAYAHG8H0.

Map 24.2 (Richest 10%): World Bank, World Development Indicators Online, 2008, http://go.worldbank.org/6HAYAHG8H0.

Map 25.1 (Gender and life expectancy): UNDP 2008, http://hdr.undp.org/en/media/HDR_20072008_EN_Indicator_tables.pdf, table 28.

Map 25.2 (Gender Development Index): UNDP 2008, http://hdr.undp.org/en/media/HDR_20072008_EN_Indicator_tables.pdf, table 28.

Map 26.1 (GNI/capita): World Bank, World Development Indicators Online, 2008, http://go.worldbank.org/6HAYAHG8H0.

Map 27.1 (Percentage of population living on $1.25/day): World Bank, PovcalNet, 2008, http://iresearch.worldbank.org/PovcalNet/povcalSvy.html.

Map 27.2 (Change in rate of extreme poverty): World Bank, PovcalNet, 2008, http://iresearch.worldbank.org/PovcalNet/povcalSvy.html.

Map 27.3 (Population in extreme poverty): World Bank, PovcalNet, 2008, http://iresearch.worldbank.org/PovcalNet/povcalSvy.html.

Map 27.4 (Change in number of people in extreme poverty): World Bank, PovcalNet, 2008, http://iresearch.worldbank.org/PovcalNet/povcalSvy.html.

Map 27.5 (Number of people living on $1.25/day): World Bank, PovcalNet, 2008, http://iresearch.worldbank.org/PovcalNet/povcalSvy.html.

Map 28.1 (Dependency ratio): World Bank, World Development Indicators Online, 2008, http://go.worldbank.org/6HAYAHG8H0.

Map 29.2 (HIV/AIDS): UNAIDS 2008, http://data.unaids.org:80/pub/GlobalReport/2008/20080820_gr08_annex1_table_en.xls.

Map 30.1 (Malaria): WHO 2008f, annex 2, 142–44, http://www.who.int/malaria/wmr2008/malaria2008.pdf.

Map 31.1 (Health expenditures): World Bank, World Development Indicators Online, 2008, http://go.worldbank.org/6HAYAHG8H0.

Map 32.1 (Safe water): World Bank, World Development Indicators Online, 2008, http://go.worldbank.org/6HAYAHG8H0.

Map 32.2 (Improved sanitation): World Bank, World Development Indicators Online, 2008, http://go.worldbank.org/6HAYAHG8H0.

Map 32.3 (Cholera epidemic): LaFraniere 2006.

Map 32.4 (Indian households with improved drinking water): IIPS and Macro International 2007, *India 2006 DHS Final Report,* http://www.measuredhs.com/pubs/pdf/FRIND3/02Chapter02.pdf, 42.

Map 33.1 (Importance of primary products): World Bank, World Development Indicators Online,, http://go.worldbank.org/6HAYAHG8H0.

Map 33.2 (Trade as a % of GDP): World Bank, World Development Indicators Online,, http://go.worldbank.org/6HAYAHG8H0.

Map 34.1 (Change in terms of trade): World Bank, World Development Indicators Online,, http://go.worldbank.org/6HAYAHG8H0.

Map 35.1 (Largest single-year decline in terms of trade): World Bank, World Development Indicators Online,, http://go.worldbank.org/6HAYAHG8H0.

Map 36.1 (Cereal trade): FAO 2008b, http://www.fao.org/statistics/yearbook/vol_1_1/site_en.asp?page=trade, table C-16.

Map 37.1 (Food price inflation): FAO 2008b, http://www.fao.org/statistics/yearbook/vol_1_1/site_en.asp?page=prices, table E-1.

Map 38.1 (Aid per capita): World Bank, World Development Indicators Online, 2008, http://go.worldbank.org/6HAYAHG8H0.

Map 38.2 (Food aid): WFP/INTERFAIS, 2008, http://www.wfp.org/interfais/index2.htm#, table 15.

Map 38.3 (Food aid by country): WFP/ INTERFAIS 2008, www.wfp.org/interfais/2008/tables/Table15.pdf.

Map 38.4 (Food aid and the extreme poor): WFP/ INTERFAIS 2008, www.wfp.org/interfais/2008/tables/Table15.pdf; World Bank, PovcalNet, 2008, http://go.worldbank.org/NT2A1XUWP0.

Map 39.1 (Military spending): World Bank, World Development Indicators Online, 2008, http://go.worldbank.org/6HAYAHG8H0. Most recent year since 2000.

Map 39.2 (Human-made disasters): World Food Programme 2006a, 146–49, B. Data Compendium, table 6.

Map 39.3 (Human-made disasters): World Food Programme 2006a, 146–49, B. Data Compendium, table 6.

Map 39.4 (Armed conflicts): Marshall and Gurr 2005, http://www.cidcm.umd.edu/pc/chapter03/.

Map 40.1 (Natural disasters): Emergency Events Database (Em-Dat), 2008, http://www.emdat.be/Database/terms.html.

Map 40.2 (Natural disasters): Emergency Events Database (Em-Dat), 2008, http://www.emdat.be/Database/terms.html.

Map 41.1 (Change in number undernourished): FAOSTAT 2008, http://www.fao.org/faostat/foodsecurity/index_en.htm.

Map 41.2 (Change in proportion undernourished): FAOSTAT 2008, http://www.fao.org/faostat/foodsecurity/index_en.htm.

# Appendix 2: Hunger Vulnerability Index

| | Available calories | Availability gap (%) | $2/day poverty rate (%) | Growth failure rate (%) | Hunger Vulnerability Index | Scaled HVI |
|---|---|---|---|---|---|---|
| Albania | 2860 | 0.00 | 7.45 | 39.20 | 15.55 | 23.17 |
| Algeria | 3040 | 0.00 | 22.96 | 22.50 | 15.15 | 19.64 |
| Angola | 2070 | 10.00 | 69.78 | 61.70 | 47.16 | 66.62 |
| Argentina | 2980 | 0.00 | 11.60 | 8.20 | 6.60 | 8.08 |
| Armenia | 2260 | 1.74 | 42.30 | 18.80 | 20.95 | 26.09 |
| Azerbaijan | 2620 | 0.00 | 0.25 | 24.10 | 8.12 | 12.62 |
| Bangladesh | 2200 | 4.35 | 80.70 | 47.80 | 44.28 | 57.44 |
| Belarus | 2960 | 0.00 | 0.47 | 4.50 | 1.66 | 2.24 |
| Benin | 2530 | 0.00 | 74.66 | 39.10 | 37.92 | 46.43 |
| Belize | 2840 | 0.00 | 27.70 | 29.10 | 18.93 | 24.81 |
| Bhutan | 2300 | 0.00 | 48.82 | 47.70 | 32.17 | 42.05 |
| Bolivia | 2220 | 3.48 | 29.98 | 32.50 | 21.99 | 30.83 |
| Botswana | 2180 | 5.22 | 48.84 | 29.10 | 27.72 | 37.27 |
| Brazil | 3060 | 0.00 | 17.97 | 13.50 | 10.49 | 13.11 |
| Bulgaria | 2850 | 0.00 | 2.26 | 8.80 | 3.69 | 5.16 |
| Burkina Faso | 2460 | 0.00 | 80.68 | 43.10 | 41.26 | 50.65 |
| Burundi | 1640 | 28.70 | 93.24 | 63.10 | 61.68 | 93.88 |
| Cameroon | 2270 | 1.30 | 56.95 | 35.40 | 31.22 | 39.59 |
| Central African Republic | 1940 | 15.65 | 81.45 | 44.60 | 47.23 | 67.10 |
| Chad | 2160 | 6.09 | 82.81 | 44.80 | 44.57 | 58.28 |
| Chile | 2860 | 0.00 | 5.13 | 6.70 | 3.94 | 5.03 |
| China | 2940 | 0.00 | 36.40 | 21.80 | 19.40 | 23.93 |
| Colombia | 2580 | 0.00 | 25.88 | 16.20 | 14.03 | 17.30 |
| Congo | 2150 | 6.52 | 73.89 | 31.20 | 37.20 | 48.36 |
| Costa Rica | 2850 | 0.00 | 8.31 | 6.10 | 4.80 | 5.82 |
| Cote d'Ivoire | 2630 | 0.00 | 46.05 | 40.10 | 28.72 | 37.04 |
| Croatia | 2770 | 0.00 | 0.00 | 0.60 | 0.20 | — |
| Czech Republic | 3240 | 0.00 | 0.00 | 2.60 | 0.87 | 1.07 |
| Democratic Republic of Congo | 1610 | 30.00 | 79.07 | 44.40 | 51.16 | 80.27 |
| Dominican Republic | 2290 | 0.43 | 14.73 | 11.70 | 8.95 | 11.46 |
| Ecuador | 2710 | 0.00 | 20.01 | 29.00 | 16.34 | 22.09 |
| Egypt | 3350 | 0.00 | 17.56 | 21.10 | 12.89 | 17.02 |
| El Salvador | 2560 | 0.00 | 24.93 | 24.60 | 16.51 | 21.45 |
| Equatorial Guinea | 2300 | 0.00 | 69.20 | 35.00 | 34.73 | 42.35 |
| Eritrea | 1520 | 33.91 | 76.50 | 43.70 | 51.37 | 82.85 |
| Ethiopia | 1860 | 19.13 | 76.72 | 50.70 | 48.85 | 72.13 |
| Gabon | 2670 | 0.00 | 19.01 | 26.30 | 15.10 | 20.30 |
| Gambia | 2280 | 0.87 | 56.04 | 27.60 | 28.17 | 34.69 |
| Georgia | 2520 | 0.00 | 29.80 | 10.90 | 13.57 | 15.83 |
| Ghana | 2650 | 0.00 | 52.86 | 35.60 | 29.49 | 37.00 |
| Guatemala | 2210 | 3.91 | 23.85 | 54.30 | 27.35 | 40.76 |
| Guinea | 2420 | 0.00 | 86.88 | 39.30 | 42.06 | 50.77 |
| Guinea-Bissau | 2070 | 10.00 | 77.25 | 36.10 | 41.12 | 55.55 |
| Guyana | 2730 | 0.00 | 16.45 | 13.80 | 10.08 | 12.75 |
| Haiti | 2090 | 9.13 | 71.69 | 29.70 | 36.84 | 49.36 |
| Honduras | 2360 | 0.00 | 34.32 | 34.50 | 22.94 | 29.98 |
| Hungary | 3500 | 0.00 | 0.00 | 3.30 | 1.10 | 1.44 |

| | Available calories | Availability gap (%) | $2/day poverty rate (%) | Growth failure rate (%) | Hunger Vulnerability Index | Scaled HVI |
|---|---|---|---|---|---|---|
| Kampuchea | 2060 | 10.43 | 67.48 | 43.70 | 40.54 | 56.65 |
| India | 2440 | 0.00 | 74.79 | 47.90 | 40.90 | 51.17 |
| Indonesia | 2880 | 0.00 | 52.84 | 28.60 | 27.15 | 33.26 |
| Iran | 3090 | 0.00 | 7.69 | 15.40 | 7.70 | 10.56 |
| Iraq | 3000 | 0.00 | 16.25* | 27.50 | 14.58 | 19.98 |
| Jamaica | 2680 | 0.00 | 5.54 | 6.60 | 4.05 | 5.12 |
| Jordan | 2680 | 0.00 | 3.24 | 12.00 | 5.08 | 7.20 |
| Kazakhstan | 2710 | 0.00 | 16.59 | 17.50 | 11.36 | 14.77 |
| Kenya | 2150 | 6.52 | 39.25 | 41.00 | 28.92 | 41.57 |
| Kyrgyzstan | 3050 | 0.00 | 50.99 | 18.10 | 23.03 | 27.02 |
| Laos | 2320 | 0.00 | 76.11 | 48.20 | 41.44 | 51.78 |
| Lebanon | 3000 | 0.00 | 11.00 | 12.20 | 7.73 | 10.00 |
| Lesotho | 2630 | 0.00 | 61.70 | 45.20 | 35.63 | 45.19 |
| Liberia | 1940 | 15.65 | 94.67 | 45.30 | 51.87 | 72.06 |
| Libya | 3330 | 0.00 | 16.25* | 20.70 | 12.32 | 16.36 |
| Macedonia | 2800 | 0.00 | 3.01 | 8.00 | 3.67 | 4.99 |
| Madagascar | 2040 | 11.30 | 89.20 | 52.80 | 51.10 | 69.89 |
| Malawi | 2140 | 6.96 | 90.15 | 52.60 | 49.90 | 65.84 |
| Malaysia | 2870 | 0.00 | 7.45 | 15.60 | 7.68 | 10.58 |
| Mali | 2230 | 3.04 | 76.50 | 38.50 | 39.35 | 49.74 |
| Mauritania | 2780 | 0.00 | 43.36 | 39.50 | 27.62 | 35.78 |
| Mexico | 3180 | 0.00 | 4.57 | 21.70 | 8.76 | 12.84 |
| Moldova | 2730 | 0.00 | 28.17 | 11.30 | 13.16 | 15.48 |
| Mongolia | 2250 | 2.17 | 48.12 | 29.80 | 26.70 | 34.40 |
| Morocco | 3070 | 0.00 | 13.43 | 23.10 | 12.18 | 16.66 |
| Mozambique | 2070 | 10.00 | 89.73 | 47.00 | 48.91 | 65.70 |
| Myanmar | 2900 | 0.00 | 73.16* | 40.60 | 37.92 | 46.71 |
| Namibia | 2260 | 1.74 | 61.88 | 29.50 | 31.04 | 38.58 |
| Nepal | 2450 | 0.00 | 77.05 | 49.30 | 42.12 | 52.70 |
| Nicaragua | 2290 | 0.43 | 31.31 | 25.20 | 18.98 | 24.41 |
| Niger | 2160 | 6.09 | 85.17 | 54.80 | 48.69 | 64.43 |
| Nigeria | 2700 | 0.00 | 83.48 | 43.00 | 42.16 | 51.57 |
| Pakistan | 2340 | 0.00 | 59.24 | 45.90 | 35.05 | 44.71 |
| Panama | 2260 | 1.74 | 17.70 | 21.50 | 13.65 | 18.99 |
| Papua New Guinea | 2250 | 2.17 | 56.80 | 50.20 | 36.39 | 48.29 |
| Paraguay | 2530 | 0.00 | 18.06 | 18.30 | 12.12 | 15.70 |
| Peru | 2570 | 0.00 | 19.02 | 31.30 | 16.77 | 22.97 |
| Philippines | 2450 | 0.00 | 44.35 | 33.80 | 26.05 | 33.09 |
| Romania | 3520 | 0.00 | 3.24 | 12.80 | 5.35 | 7.63 |
| Russia | 3080 | 0.00 | 1.39 | 12.70 | 4.70 | 6.94 |
| Rwanda | 2070 | 10.00 | 90.02 | 51.70 | 50.57 | 68.30 |
| Senegal | 2310 | 0.00 | 59.57 | 20.10 | 26.56 | 31.06 |
| Sierra Leone | 1930 | 16.09 | 75.53 | 38.40 | 43.34 | 62.17 |
| South Africa | 2940 | 0.00 | 42.42 | 22.80 | 21.74 | 26.55 |
| Sri Lanka | 2390 | 0.00 | 38.90 | 18.40 | 19.10 | 22.98 |
| Sudan | 2260 | 1.74 | 72.44* | 47.50 | 40.56 | 51.85 |
| Suriname | 2660 | 0.00 | 26.80 | 14.50 | 13.77 | 16.71 |
| Swaziland | 2360 | 0.00 | 80.59 | 36.60 | 39.06 | 47.15 |
| Syria | 3060 | 0.00 | 16.25* | 18.80 | 11.68 | 15.34 |
| Tajikistan | 1840 | 20.00 | 49.90 | 33.10 | 34.33 | 54.30 |
| Tanzania | 1960 | 14.78 | 96.11 | 44.40 | 51.76 | 71.22 |
| Thailand | 2410 | 0.00 | 11.02 | 15.70 | 8.91 | 11.88 |
| Togo | 2320 | 0.00 | 68.51 | 29.80 | 32.77 | 39.33 |
| Trinidad and Tobago | 2770 | 0.00 | 13.15 | 5.30 | 6.15 | 7.07 |

| | Available calories | Availability gap (%) | $2/day poverty rate (%) | Growth failure rate (%) | Hunger Vulnerability Index | Scaled HVI |
|---|---|---|---|---|---|---|
| Tunisia | 3250 | 0.00 | 12.38 | 8.30 | 6.89 | 8.40 |
| Turkey | 3340 | 0.00 | 8.77 | 15.60 | 8.12 | 11.04 |
| Turkmenistan | 2750 | 0.00 | 48.87 | 22.30 | 23.72 | 28.52 |
| Uganda | 2380 | 0.00 | 75.02 | 44.80 | 39.94 | 49.59 |
| Ukraine | 3030 | 0.00 | 0.46 | 22.90 | 7.79 | 12.05 |
| Uruguay | 2850 | 0.00 | 1.80 | 13.90 | 5.23 | 7.72 |
| Uzbekistan | 2270 | 1.30 | 75.98 | 25.30 | 34.19 | 40.81 |
| Venezuela | 2350 | 0.00 | 31.19 | 14.60 | 15.26 | 18.28 |
| Vietnam | 2580 | 0.00 | 47.57 | 43.40 | 30.32 | 39.33 |
| Yemen | 2020 | 12.17 | 45.65 | 51.70 | 36.51 | 55.05 |
| Zambia | 1930 | 16.09 | 81.11 | 52.50 | 49.90 | 71.62 |
| Zimbabwe | 2010 | 12.61 | 72.44* | 35.80 | 40.28 | 56.29 |

Notes: Available calories are measured per person per day for 2001 through 2003. Availability gap is the percentage by which caloric availability falls below 2300. The poverty rate is calculated using purchasing power parity conversions for 2005. Growth failure rate is for children under five years of age as of 2005. Hunger Vulnerability Index is the average of the availability gap, $2/day poverty rate, and rate of growth failure. The scaled HVI is calculated by scaling each component to range from 0 to 100 before averaging.

*Indicates use of regional average rather than national data.

Sources: Availability data from FAO 2008a . Poverty data from PovcalNet, World Bank 2009. Growth failure data are from WHO 2004c.

For the following countries and regions there are no reliable estimates of the $2/day poverty rate. Rates of growth failure and availability provided.

| | Availability gap (%) | Growth failure rate (%) |
|---|---|---|
| Afghanistan | N/A | 59.30 |
| Bahrain | N/A | 9.70 |
| Barbados | 0 | 10.20 |
| Bosnia and Herzegovina | 0 | 11.80 |
| Cape Verde | 0 | 15.30 |
| Comoros | 23.91 | 46.90 |
| Djibouti | N/A | 25.70 |
| Dominica | 0 | 9.80 |
| Qatar | N/A | 8.10 |
| St. Lucia | 0 | 10.80 |
| Samoa | 0 | 1.90 |
| São Tomé and Principe | 0 | 35.20 |
| Serbia | 0 | 8.10 |
| Seychelles | 0 | 5.10 |
| Tonga | N/A | 1.30 |
| Vanuatu | 0 | 19.10 |
| West Bank | N/A | 6.90 |

Sources: See table above.

Note: N/A indicates no estimate available for 2001–3.

For the following countries and regions there are no reliable estimates of the $2/day poverty rate, nor the rate of growth failure. In all cases the availability gap is zero or unavailable.

| | | |
|---|---|---|
| Australia | Ireland | Portugal |
| Austria | Italy | Saudi Arabia |
| Belgium | Kuwait | Singapore |
| Brunei | Latvia | Slovakia |
| Cuba | Lithuania | Slovenia |
| Denmark | Luxembourg | Somalia |
| Estonia | Malta | South Korea |
| Fiji | Mauritius | Spain |
| Finland | Netherlands | Sweden |
| France | New Zealand | Switzerland |
| Germany | North Korea | United Arab Emirates |
| Greece | Oman | United Kingdom |
| Iceland | Poland | United States |

# Notes

## Chapter One

1. Personal communication, March 2007.

2. From *Unnatural Causes: In Sickness and in Wealth,* episode 1, California Newsreel, 2008.

3. Cornell University, Moderate malnutrition kills millions of children needlessly, *Science Daily,* July 1, 2003, http://www.sciencedaily.com/releases/2003/06/030630110813.htm (accessed July 17, 2008).

4. Figure 1.2 is similar to the conceptual framework for the causes of malnutrition presented in UNICEF (1990).

5. According to the World Bank's World Development Indicators, 17.6% of Mexicans live below the national poverty line.

6. The 2300-calorie limit is widely used (Leslie et al., 1997).

7. http://www.fao.org/es/ess/fbsforte.asp.

8. Svedberg (2001) and Smith (1998) provide information on the method used for assessing distribution in the POU calculations. Svedberg argues that there is overestimation of hunger, while Smith suggests just the opposite.

9. Many other indicators of hunger provide only an index of intensity that cannot be translated into numbers of people.

10. Detailed information on the Demographic and Health Surveys can be found at http://www.measuredhs.com.

11. The international poverty line was set at $1.08 in 1993 prices, and is frequently referred to as the "dollar-a-day poverty line." In 2008 data were released reflecting increases in the cost of living and yielding a threshold for extreme poverty of $1.25 in 2005 prices. Both poverty lines attempt to capture the same quantity of consumption in terms of goods and services.

12. http://www.stephenlewisfoundation.org/news_speech_item.cfm?news=1114&year=2003.

# References

Africa Rice Center. [N.d.]. http://www.warda.org.

Algert, A., A. Agrawal, and D. Lewis. 2006. Disparities in access to fresh produce in low-income neighborhoods in Los Angeles. *American Journal of Preventive Medicine* 30 (5): 365–70.

Allen, J. A. 1998. *Student Atlas of World Politics.* 3rd ed. Guilford, CT: Dushkin/McGraw-Hill.

ANEP (Administración Nacional de Educación Pública), Programa de Alimentación Escolar. 2003. *Tercer censo nacional de talla en niños de primer grado escolar.* Montevideo: ANEP.

Bagchi, A. K. 1983. *The Political Economy of Underdevelopment.* Cambridge: Cambridge University Press.

Barquera, S., C. Hotz, J. Rivera, L. Tolentino, J. Espinoza, I. Campos, and T. Shamah. 2006. Food consumption, food expenditure, anthropometric status and nutrition-related diseases in Mexico. In *The Double Burden of Diseases in Developing Countries,* FAO Food and Nutrition Paper 84, 161–203. Rome: Food and Agricultural Organization. http://www.fao.org/docrep/009/a0442e/a0442e0k.htm#TopOfPage.

Barrett, C., and D. Maxwell. 2005. *Food Aid after Fifty Years: Recasting Its Role.* London: Routledge.

BBC (British Broadcasting Corporation). 2006. Burkina's white gold fails to deliver wealth. *BBC News,* July 25. http://news.bbc.co.uk/1/hi/business/5213562.stm.

———. 2008. UN increases food aid by $1.2 bn. *BBC News,* June 5. http://news.bbc.co.uk/2/hi/europe/7435265.stm.

Benson, T. 2004. *Assessing Africa's food and nutrition security situation.* IFPRI 2020 Africa Conference Policy Brief. Washington, DC: International Food Policy Research Institute.

Bhattacharya, D. 2003. Final countdown to the MFA: Fallout for the LDCs. Dhaka: Centre for Policy Dialogue.

Black, R. E., et al. 2008. Maternal and child undernutrition: Global and regional exposures and health consequences. *Lancet* 371(9608): 243–60.

Blades, D. W. 1980. What do we know about levels and growth of output in developing countries? A critical analysis with special reference to Africa. In *Economic Growth and Resources,* vol. 2, *Trends and Flows,* edited by R. C. O. Matthews, 60–77. London: Macmillan Press.

Boko, M., et al. 2007. *Africa Climate Change 2007: Impacts, Adaptation and Vulnerability; Contribution of Working Group II to the Fourth Assessment Report of the Intergovernmental Panel on Climate Change,* edited by M. L. Parry et al. UL 433–67. Cambridge: Cambridge University Press.

Bradsher, K. 2008. A new global oil quandary: Costly fuel means costly calories. *New York Times,* January 19.

Bread for the World. 2006. *Hunger Report 2006: Frontline Issues in Nutrition Assistance.* Washington, DC: Bread for the World Institute.

Brown, L., H. Feldstein, L. Haddad, C. Peña, and A. Quisumbing. 1995. Generating food security in the year 2020: Women as producers, gatekeepers, and shock absorbers. IFPRI 2020 Vision Brief 17. Washington, DC: International Food Policy Research Institute.

Carroll, R. 2007. Venezuela scrambles for food despite oil boom. *Guardian,* November 14.

CDC (Centers for Disease Control and Prevention). 2008. Overweight and obese: U.S. obesity trends 1985–2006. http://www.cdc.gov/nccdphp/dnpa/obesity/trend/maps/ (accessed on January 23, 2008).

Cohen, R., M. Kim, and J. Ohls. 2006. *Hunger in America 2006: National Report Prepared for America's Second Harvest.* Princeton, NJ: Mathematica Policy Research.

Congressional Research Service. 2008. *International Food Aid Provisions of the 2008 Farm Bill: CRS Report to Congress.* Order code RS22900. http://www.nationalaglawcenter.org/assets/crs/RL34145.pdf.

de Onis, M., C. Garza, A. Onyango, and E. Borghi. 2007. Comparison of the WHO child growth standards and the CDC 2000 growth charts. *Journal of Nutrition* 137: 144–48.

de Waal, A. 1997. *Famine Crimes: Politics and the Disaster Relief Industry.* London: James Currey and Indiana University Press.

Digger, C. 2007. A bounty of rice for Africa, just out reach. *New York Times,* October 10.

Dixon, J., A. Gulliver, and D. Gibbon. 2001. *Farming Systems and Poverty: Improving Farmers' Livelihoods in a Changing World.* Rome: FAO/World Bank.

Drewnowski, A., and S. E. Spencer. 2004. Poverty and obesity: The role of energy density and energy costs. *American Journal of Clinical Nutrition* 79: 6–16.

Easterling, W., et al. 2007. Food, fibre and forest products. In *Climate Change 2007: Impacts, Adaptation and Vulnerability; Contribution of Working Group II to the Fourth Assessment Report of the Intergovernmental Panel on Climate Change,* edited by M. L. Parry et al. UL 273–313. Cambridge: Cambridge University Press.

Em Dat (Emergency Events Database). N.d. http://www.emdat.be/ (accessed September 12, 2008).

Esty, D. C., M. A. Levy, T. Srebotnjak, and A. de Sherbinin. 2005. *2005 Environmental Sustainability Index: Benchmarking National Environmental Stewardship.* New Haven, CT: Yale Center for Environmental Law and Policy.

Estur, G. 2005. Is West Africa competitive with the U.S. on the world cotton market? Paper presented to the Beltwide Cotton Economics and Marketing Conference, New Orleans, January 7.

FAO (Food and Agriculture Organization of the United Nations). 1996. *The Sixth World Food Survey.* Rome: FAO.

————. 1998. Food balance sheets and food consumption surveys: A comparison of methodologies and results. http://www.fao .org/es/ess/consweb.asp.

————. 2004a. *The State of Food Insecurity in the World 2004.* Rome: FAO.

————. 2004b. Cotton: Impact of support policies on developing countries—why do the numbers vary? FAO Trade Policy Brief 1. Rome: FAO.

————. 2006a. The nutrition transition and obesity. http://www .fao.org/FOCUS/E/obesity/obes2.htm (accessed on September 19, 2006).

————. 2006b. *The State of Food Insecurity in the World 2006.* Rome: FAO.

————. 2007a. *FAOSTAT.* http://faostat.fao.org/site/368/default .aspx (accessed January 2007).

————. 2007b. *OECD-FAO Outlook: 2007–2016.* Paris: OECD-FAO.

————. 2007c. *The State of Food and Agriculture: Food Aid for Food Security?* Rome: FAO.

————. 2008a. *FAOSTAT.* http://faostat.fao.org (accessed September 4, 2008).

————. 2008b. *FAO Statistical Yearbook 2005–2006.*

————. 2008c. *The State of Food Insecurity in the World 2008.* Rome: FAO. http://www.fao.org/statistics/yearbook/vol_1_1/pdf/ d01.pdf.

Feeding America. 2008. New USDA statistics highlight growing hunger crisis in the US. Press release, November 17, 2008. http://feedingamerica.org/newsroom/press-release-archive/ usda-hunger-statistics.aspx (accessed November 19, 2008).

Ferro-Luzzi, A., S. Morris, S. Taffesse, T. Demissie, and M. D'Amato. 2001. *Seasonal undernutrition in Ethiopia: Its magnitude, correlations and functional significance.* IFPRI Research Report 118. Washington, DC: International Food Policy Research Institute.

Finnegan, W. 2002. Letter from Bolivia: Leasing the rain. *The New Yorker,* April 8.

Fischer, G., M. Shah, F. Tubiello, and H. van Vehluizen. 2005. Socio-economic and climate change impacts on agriculture: An integrated assessment, 1990–2080. *Philosophical Transactions of the Royal Society B* 360: 2067–83.

FLO (Fairtrade Labeling Organization). 2007. Global fairtrade sales increase by 40%, Benefit 1.4 million farmers worldwide. FLO press release, August 10.

Freedom House. 2006. *Freedom in the World: The Annual Survey of Political Rights and Civil Liberties.* http://www.freedomhouse .org/template.cfm?page=35&year=2006.

————. 2008. *Freedom in the World 2008.* http://www.freedom house.org/template.cfm?page=414.

Galbraith, J. K. 2004. Rising inequality within countries under globalization. Paper presented at the annual meetings of the American Political Science Association. September. Chicago, IL.

Gentilini, U., and Webb, P. 2005. How are we doing on poverty and hunger reduction? A new measure of country-level progress. *Food Policy and Applied Nutrition Program.* Discussion Paper 31. Friedman School of Nutrition Science and Policy, Tufts University. http://nutrition.tufts.edu/docs/pdf/fpan/ wp31-poverty_hunger_reduction.pdf.

Global Fund to Fight AIDS, Tuberculosis and Malaria. N.d. *The Disease Report.* http://www.theglobalfund.org/en/about/ malaria/default.asp (accessed October 26, 2008).

Golden Rice Project. 2006. Biofortified rice, a contribution to alleviation of life-threatening micronutrient deficiencies in developing countries. www.goldenrice.org.

Gresser, C., and Sophia Tickell. 2002. *Mugged: Poverty in Your Cup.* Washington, DC: Oxfam International.

Gulliver, A., 2001. Private sector-led diversification among indigenous producers in Guatemala. In *Global Farming Systems Study: Challenges and Priorities to 2030,* edited by J. Dixon, A. Gulliver, and D. Gibbon. Consultation Documents. Washington, DC: World Bank/FAO.

Harley, J. B. 1989. Deconstructing the map. *Cartographica* 26: 1–20.

HarvestPlus. 2003. Micronutrient Malnutrition. http://www .harvestplus.org/micronut.html.

Hitz, S., and J. Smith. 2004. Estimating global impacts from climate change. *Global Environmental Change* 14: 201–18.

Hochschild, A. 1999. *King Leopold's Ghost.* New York: Mariner Books.

IIPS (International Institute for Population Sciences) and Macro International. 2007. *National Family Health Survey (NFHS-3), 2005–06: India,* vol. 1. Mumbai: IIPS.

IPCC (Intergovernmental Panel on Climate Change). 2007. *Intergovernmental Panel on Climate Change Fourth Assessment Report: Climate Change 2007; Synthesis Report; Summary for Policymakers.* http://www.ipcc.ch/pdf/assessment-report/ar4/syr/ ar4_syr_spm.pdf.

Jaffee, D. 2007. *Brewing Justice: Fair Trade Coffee, Sustainability, and Survival.* Berkeley: University of California Press.

Kennedy, E., V. Mannar, and V. Iyengar. 2003. Alleviating hidden hunger: Approaches that work. *International Atomic Energy Agency (IAEA) Bulletin* 45 (1): 54–60.

Kennedy, E., and P. Peters. 1992. Household food security and child nutrition: The interaction of income and gender of household head. *World Development* 20 (8): 1077–1085.

Kirby, A. 2003. "Mirage of GM's Golden Promise." *BBC News.* September 24. http://news.bbc.co.uk/go/pr/fr/-/2/hi/ science/nature/3122923.stm.

Lacey, M. 2005. The mournful math of Darfur: The dead don't add up. *New York Times,* May 18.

————. 2008. Across the globe, empty bellies bring rising anger. *New York Times,* April 18.

LaFraniere, S. 2006. In Oil-rich Angola: Cholera preys upon poorest. *New York Times,* June 16.

Leslie, J., E. Ciemins, and S. Bibi Essma. 1997. Female nutritional status across the life-span in sub-Saharan Africa: 1. Prevalence patterns. *Food and Nutrition Bulletin* 18 (1): 20–43.

Lobina, E. 2000. Cochabamba-water war. *Focus (PSI Journal)* 7:2.

Logan, J. 2007. The impact of Katrina: Race and class in storm-damaged neighborhoods. http://www.s4.brown.edu/Katrina/ report.pdf.

MacKinnon, D. 2006. Government whitewashes hunger. *Chicago Tribune,* December 1.

McGreal, C. 2007. Zimbabwe runs out of bread. *Guardian,* October 1.

McKenzie, D., and I. Ray. 2005. Household water delivery options in urban and rural India. Stanford Center for International Development Working Paper 224, December. http://scid.stanford.edu/events/India2004/McKenzie-Ray%205-11-04.pdf.

McKinley, J. C. 2007. Cost of corn soars, forcing Mexico to set price limits. *New York Times,* January 19.

Malkin, E. 2007. Thousands in Mexico City protest rising food price. *New York Times,* February 1.

Malthus, T. 1985 (1798). *An Essay on the Principle of Population.* London: Penguin.

Marshall, M. G., and T. R. Gurr. 2005. *Peace and Conflict 2005.* College Park, MD: Center for International Development and Conflict Management. http://www.cidcm.umd.edu/pc/chapter03/.

Martorell, R. 2006. Obesity: An emerging health and nutrition issue in developing countries. In *The Double Burden of Diseases in Developing Countries,* 49–53. FAO Food and Nutrition Paper 84, 49–53. Rome: Food and Agricultural Organization.

Minot, N., and Daniels, L. 2005. Impact of global cotton markets on rural poverty in Benin. *Agricultural Economics* 33, supplement, 453–66.

Mtika, M. M. 2001. The AIDS epidemic in Malawi and its threat to household food security. *Human Organization* 60 (2): 178–88.

Murphy, S., and K. McAfee. 2005. *U.S. Food Aid: Time to Get It Right.* Minneapolis, MN: Institute for Agriculture and Trade Policy.

Naiken, L. 2003. FAO methodology for estimating the prevalence of undernourishment. In *Measurement and Assessment of Food Deprivation and Undernutrition.* Proceedings of an International Scientific Symposium convened by the Agriculture and Economic Development Analysis Division of the FAO, 7–42. FAO: Rome.

National Archives of France, Overseas Section. Photothéque No. 2220. "Tiébissou—marché du coton."

Naylor, R., and W. Falcon. 2008. Our daily bread: Without public investment, the food crisis will only get worse. *Boston Review* 33 (5): 13–18.

Nord, M., M. Andrews, and S. Carlson. 2007. *Household Food Security in the United States, 2006.* USDA, Economic Research Service, Economic Research Report 49, Washington, DC. http://www.ers.usda.gov/Publications/ERR49/ERR49appD.pdf.

OECD (Organization for Economic Co-operation and Development). 2005. *Economic and Social Importance of Cotton Production and Trade in West Africa: Role of Cotton in Regional Development, Trade, and Livelihoods.* Paris: Sahel and West Africa Club Secretariat/OECD.

———. 2007. *Geographical Distribution of Financial Flows to Aid Recipients 2001/2005.* Paris: OECD.

O'Loughlin, J. 2005. The political geography of conflict: Civil wars in the hegemonic shadow. In *The Geography of War and Peace: From Death Camps to Diplomats,* edited by C. Flint, 85–110. Oxford: Oxford University Press.

Oxfam. 2002. *Cultivating Poverty: The Impact of US Cotton Subsidies on Africa.* Oxfam Briefing Paper #30.

———. 2008. *The Time Is Now: How World Leaders Should Respond to the Food Price Crisis.* Oxfam Briefing Note, June.

Parry, M., C. Rosenzweig, A. Inglesias, M. Livermore, and G. Fischer. 2004. Effects of climate change on global food production under SRES emissions and socio-economic scenarios. *Global Environmental Change* 14: 53–67.

Pearson, N. O. 2007. Meat, sugar scarce in Venezuela stores. *Washington Post,* February 8.

Pedro, M. R. A., and R. C. Benavides. 2006. Dietary changes and their health implications in the Philippines. In *The Double Burden of Diseases in Developing Countries,* FAO Food and Nutrition Paper 84. Rome: Food and Agricultural Organization.

Phelan, J. C., and B. Link. 2005. Controlling disease and creating disparities: A fundamental cause perspective. *Journal of Gerontology,* ser. B, vol. 60B (Special Issue II): 27–33.

Ponte, S. 2002. The "Latte Revolution"? Regulation, markets, and consumption in the global coffee chain. *World Development* 30 (7): 1099–1122.

Porter. E. 2006. Flows of immigrants' money to Latin America surges. *New York Times,* October 14.

Ravallion, M., S. Chen, and P. Sangraula. 2008. Dollar a day revisted. Policy Research Working Paper 4620. Washington DC: World Bank.

Renewable Fuels Association. 2007. http://www.ethanolrfa.org/industry/statistics/#A.

Ribot, J. 1996. Climate variability, climate change and vulnerability: Moving forward by looking back. Introduction to *Climate Variability, Climate Change and Social Vulnerability in the Semiarid Tropics,* edited by J. Ribot, 1–10. Cambridge: Cambridge University Press.

Rivoli, P. 2005. *The Travels of a T-Shirt in the Global Economy.* Hoboken, NJ: John Wiley & Sons.

Rohter, L. 2005. Brazil, too, weighs in on obesity. *New York Times,* January 14.

Rosenthal, E. 2007. World food supply is shrinking, UN agency warns. *New York Times* December 18.

———. 2008. UN says biofuel subsidies raise food bill and hunger. *New York Times,* October 8.

Sachs, J. 2003. Institutions matter, but not for everything. *Finance and Development* 40 (2): 38–41.

———. 2005. The development challenge. *Foreign Affairs* 84 (2): 78–90.

Sanchez, P., M. S. Swaminathan, P. Dobie, and N. Yuksel. 2005. *Halving Hunger: It Can Be Done.* Summary version. UN Millenium Project Task Force on Hunger.

Schrader, F., and L. Gallouédec. 1914. *Atlas classique de géographie ancienne et moderne.* Paris: Librairie Hachette.

Scott, J. 1976. *The Moral Economy of the Peasant.* New Haven: Yale University Press.

Sen, A. 1981. *Poverty and Famines: An Essay on Entitlement and Deprivation.* New York: Oxford University Press.

———. 1990. Individual freedom as a social commitment. *New York Review of Books* 37 (10) (June 14).

Shetty, P. 2003. Measures of nutritional status from anthropometric survey data. In *Measurement and Assessment of Food Deprivation and Undernutrition,* edited by FIVIMS, 139–61. Rome: FAO.

Shultz, J. 2005. The politics of water in Bolivia. *Nation,* online article posted on January 28. http://www.thenation.com/doc/20050214/shultz.

Smith, L. 1998. *Can the FAO's Measure of Chronic Undernourishment Be Strengthened?* (With response by Logenden Naiken). Food Consumption and Nutrition Division Discussion Paper 44. Washington, DC: International Food Policy Research Institute.

Smith, L., U. Ramakrishnan, A. Ndiaye, L. Haddad, and R. Martorell. 2003. *The Importance of Women's Status for Child Nutrition in Developing Countries.* IFPRI Research Report 131. Washington, DC: IFPRI.

Stiglitz, J. 2003. *Globalization and Its Discontents.* New York: W. W. Norton.

Svedberg, P. 1999. 841 million undernourished? *World Development* 27 (12): 2081–2098.

———. 2001. Undernutrition overestimated. Institute for International Economic Studies, Stockholm University, Seminar Paper 693. October.

Swaminathan Research Foundation and the World Food Program. 2001. *Food Insecurity Atlas of Rural India.* Chennai: Swaminathan Research Foundation.

Thurow, R., and S. Kilman. 2003. As U.S. food aid enriches farmers, poor nations cry foul. *The Wall Street Journal,* September 11.

———. 2005. Farmers, charities join forces to block famine-relief revamp. *The Wall Street Journal,* October 26.

Tubiello, F. N., and G. Fischer. 2007. Reducing climate change impacts on agriculture: Global and regional effects of mitigation, 2000–2080. *Technological Forecasting and Social Change* 74: 1030–56.

UNAIDS. 2006. *Report on the Global Aids Epidemic: A UNAIDS 10th Anniversary* Special Edition. Geneva: UNAIDS.

———. 2008. *2008 Report on the Global AIDS Epidemic.* Geneva: UNAIDS. http://www.unaids.org/en/KnowledgeCentre/HIVData/GlobalReport/2008/2008_Global_report.asp (accessed October 23, 2008).

UNDP (United Nations Development Program). 2002. *Human Development Report 2002: Deepening Democracy in a Fragmented World.* New York: Oxford University Press.

———. 2005. *Human Development Report 2005: International Cooperation at a Crossroads; Aid, Trade and Security in an Unequal World.* New York: Oxford University Press.

———. 2008. *Human Development Report 2007–2008.* http://hdr.undp.org/en/media/HDR_20072008_EN_Indicator_tables.pdf.

UNICEF (United Nations Children's Fund). 1990. *Strategy for Improved Nutrition of Children and Women in Developing Countries.* New York: UNICEF.

———. 1998. *State of the World's Children.* Oxford University Press: Oxford.

———. 2007. UNICEF Statistics. http://www.childinfo.org/sanitation.html (accessed October 27, 2008).

United Nations Standing Committee on Nutrition. 1997. *Third Report on the World Nutrition Situation.* Geneva: United Nations.

———. 2004. *Fifth Report on the World Nutrition Situation.* http://www.unsystem.org/scn/publications/AnnualMeeting/SCN31/SCN5Report.pdf (accessed October 20, 2008).

United States Department of Agriculture. 2008. *Food Security Assessment, 2007.* Washington, DC. Economic Research Services/USDA.

van Jaarsveld, P. J., M. Faber, S. Tanumihardjo, P. Nestel, C. J. Lombard, and A. J. Spinnler Benadé. 2005. ß-Carotene–rich orange-fleshed sweet potato improves the vitamin A status of primary school children assessed with the modified-relative-dose-response test. *American Journal of Clinical Nutrition* 81 (5): 1080–1087.

Via Campesina. 2008. FAO opening: *Via Campesina* farmers to the heads of state: Time to change food policies. Press Release, June 3.

Von Braun, J. 2007. *The World Food Situation: New Driving Forces and Required Actions.* Washington, DC: IFPRI.

Von Grebmer, K., H. Fritschel, B. Nestorova, T. Olofinbiyi, R. Pandya-Lorch, and Y. Yohannes. 2008. *The Challenge of Hunger: The 2008 Global Hunger Index.* IFPRI Issue Brief 54. Washington, DC: IFPRI.

Watts, M. 1983. *Silent Violence: Food, Famine and Peasantry in Northern Nigeria.* Berkeley: University of California Press.

———. 1987. Drought, environment and food security: Some reflections on peasants, pastoralists, and commodification in dryland West Africa. In *Drought and Hunger in Africa,* edited by M. Glantz, 71–211. Cambridge: Cambridge University Press.

Watts, M., and H. Bohle. 1993. The space of vulnerability: The causal structure of hunger and famine. *Progress in Human Geography* 17 (1): 43–68.

Weiner, D., S. Moyo, B. Munslow, and P. O'Keefe. 1985. Land use and agricultural productivity in Zimbabwe. *Journal of Modern African Studies* 23 (2): 251–85.

Weismann, D. 2006. *Global Hunger Index: A Basis for Cross-Country Comparisons.* Washington, DC: IFPRI.

WHO (World Health Organization of the United Nations ). 2004a. *Fifth Report on the World Nutrition Situation.* Geneva: WHO.

———. 2004b. *Iodine Status Worldwide.* WHO global database on iodine deficiency. Geneva: WHO. http://www.who.int/vmnis/iodine/data/en/index.html (accessed on October 27, 2008).

———. 2004c. WHO Global Database on Child Growth and Malnutrition. http://www.who.int/nutgrowthdb/database/en/ (accessed on October 27, 2008).

———. 2006a. Obesity and Overweight. Global Strategy on Diet, Physical Activity and Health. http://www.who.int/dietphysicalactivity/en/ (accessed on October 27, 2008).

———. 2006b. World Health Organization releases new child growth standards. http://www.who.int/mediacentre/news/

releases/2006/pr21/en/index.html (accessed on October 27, 2008).

———. 2008c. *Global Database on Child Health and Malnutrition.* http://www.who.int/nutgrowthdb/database/countries/en/index.html (accessed on October 27, 2008).

———. 2008d. Micronutrient Deficiency Information System. http://www.who.int/nutrition/databases/micronutrients/en/print.html.

———. 2008e. *Statistical Information System.* http://www.who .int/whosis/en/ (accessed on October 27, 2008).

———. 2008f. *World Malaria Report 2008.* http://malaria.who.int/wmr2008/ (accessed October 22, 2008).

———. 2008g. *Worldwide Prevalence of Anaemia, 1993–2005.* WHO Global Database on Anaemia. Geneva: WHO. http://whqlibdoc.who.int/publications/2008/9789241596657_eng.pdf (accessed on October 27, 2008).

———. 2008h. *Vitamin and Mineral Nutrition Information System.* http://www.who.int/vmnis/en/index.html.

WHO/UNICEF/ICCIDD. 2001. *Assessment of Iodine Deficiency Disorders and Monitoring Their Elimination: A Guide for Programme Managers,* 2nd ed. Geneva: WHO.

Wines, M. 2005. Niger's anguish is reflected in its dying children. *New York Times* August 5.

Wolford, W. 2004. This land is ours now: Spatial imaginaries and the struggle for land in Brazil. *Annals of the Association of American Geographers* 94 (2): 409–24.

Wood, D. 1992. *The Power of Maps.* New York: Guilford Press.

World Bank. 2006. *Repositioning Nutrition as Central to Development: A Strategy for Large-scale Action.* Washington, DC: World Bank.

———. 2008a. *World Development Indicators.* Washington DC: World Bank.

———. 2008b. *World Development Report 2008: Agriculture for Development.* Washington DC: World Bank.

———. N.d. PovcalNet. http://go.worldbank.org/NT2A1XUWP0 (accessed September 2008).

———. N.d. *World Development Indicators Online.* http://web .worldbank.org/WBSITE/EXTERNAL/DATASTATISTICS/0,,contentMDK:20398986~menuPK:64133163~pagePK:64133150~piPK:64133175~theSitePK:239419,00.html (accessed September 2008).

World Food Programme. 2006a. *World Hunger Series 2005: Hunger and Learning.* Rome: World Food Programme and Stanford University Press.

World Food Programme. 2006b. *Food Aid Information System.* http://www.wfp.org/fais/.

Wright, A., and W. Wolford. 2003. *To Inherit the Earth: The Landless Movement and the Struggle for a New Brazil.* Oakland, CA: Food First Books.

Ziegler, J. B., and R. A. Ffrench. 2000. Nelson Mandela argues for urgent action against HIV in Africa. Conference Report. XIII International AIDS Conference, Durban, July 9–14, 2000. *Medical Journal of Australia* 173: 572–74. http://www.mja.com .au/public/issues/173_11_041200/ziegler/ziegler.html#box4 (accessed August 29, 2009).

# Index

adjusted savings, 91

Africa: biofortification, 27; climate change, 94, 95; coffee, 160, 161; colonialism, 109, 110; cotton, 158; fertilizer, 103, 105; GNI, 128; growth failure 37, 40; health expenditures, 145; HIV/AIDS, 139, 140; HVI, 56; iron deficiency, 30; malaria, 143; NERICA, 104; obesity, 33; population, 76; poverty, 45, 46, 49, 131; trade, 156; undernourishment 19, 22, 25, 182; war in, 174; women in, 124

Africa Rice Center, 104

agricultural subsidies, 158

agricultural technology, 97

anemia: in India, 67

Angola: unsafe water in, 149–50

anthropometric indicators: advantages of, 40–41; limitations of, 41

arable land: changes in, 82; and hunger vulnerability, 79–83

Bangladesh: poverty in, 134

biofortification: and micronutrient deficiencies, 27–28; and Golden Rice, 28; and Vitamin A deficiency 29; and iodine deficiency, 30–31

biofuels, 163–64

Body Mass Index (BMI), 32–33

Bolivia: water privatization in, 153

Botswana: HIV/AIDS in, 141

Brazil: hunger in, 19; land distribution in, 80; Landless Workers Movement, 80; Zero Hunger Project, 5, 181

caloric cutoff: and POU hunger indicator, 19

caloric requirements: minimum daily needs, 20

carbon dioxide emissions, 78, 94

Central America: growth failure, 37; GNI 128

China: food availability in, 16; health care spending in, 146; hunger in, 19; poverty in, 131

cholera: in Angola, 149–50

climate change, 94–95

coffee crisis, 161

colonialism, 109–11

conflicts: and hunger, 174–77

consumption pressure, 78

cotton: and hunger vulnerability, 158–59; and trade, 155

debt: crisis, 112; and hunger, 112–14; and the World Bank, 112; and the IMF, 112

Demographic and Health Surveys, 30

dependency ratio: definition of, 137; composition of dependents, 138; and HIV/AIDS, 138–39, and hunger, 137–39

development aid: and food security, 168–70; and the United States, 168

East Asia: poverty, 45, 46, 131; undernourishment, 22

education: higher, 99

entitlements: definition of, 5; failure, 6; programs to reduce hunger, 181

environmental systems, 84–85, 184

ethanol, 8

Ethiopia: hunger in, 21

Europe: biofuels, 163, 164; colonialism, 109–11; obesity, 33

Feeding America, 66

fertilizer, 103–5

food: national availability of 2–4; costs as share of household budget, 165; sovereignty, 9; trade in, 162–66

food aid: controversy in the United States, 169; and hunger, 168–73; to people living in poverty, 173; recipients of, 172; targeting, 171; types of, 170–71

Food and Agricultural Organization (FAO): and global food crisis 2; and food trade, 162; and healthy diets, 17; POU hunger indicator, 19–20

food availability: errors in calculating, 19–21; and food balance sheets, 15–16; global, 163; in Hunger Vulnerability Index, 51; as indicator of hunger, 25; as indicator of malnutrition and hunger, 13, 15–18; as measured by FAO, 15–17

food insecurity: definition of, 3; and food price inflation, 1–2; in the United States, 63–65

food price: dilemma, 167; in Mexico, 165; inflation and undernutrition, 166

food price shocks: and growth failure, 165; and hunger, 165–67; and natural disasters, 178

food protests: in Haiti, 1; in Mexico, 6–7; riots, 164

food security: definition of, 3; and development aid, 168–70; household, 122; and national food availability, 4; and women, 122–24

food stamps, in the United States, 5

Freedom House, 116

gender inequality, and hunger vulnerability, 120–24, 184

Gender-related Development Index (GDI), 122–23

Gini coefficient, 119

Global Hunger Index, 54

gross national income (GNI), problems in using to locate hunger, 127

growth charts, 38

growth failure: definition of, 38; and hunger, 37–42; in the Hunger Vulnerability Index, 52; in India, 68; in Montevideo, 61–62; urban-rural differences in, 60–61; in Uruguay, 57–62

Haiti, hunger protests in, 1

health care: in China, 146, expenditures and hunger vulnerability, 145–46, 184

Highly Indebted Poor Countries (HIPC) Initiative, 113

HIV/AIDS: in Botswana, 141; and dependency ratio, 138–39; and food security, 140–42; and orphans, 141–42; in Zimbabwe, 78

hunger: and colonialism, 109–11; within countries, 57–69; definition of, 2–3; and health spending, 146; and international trade, 154–56; and neocolonialism, 111; and power relations, 107; protests, 1, 4, 164; seasonal, 21; sources of, 5–6, 71, 181–84; and terms of trade, 157–61; and trade in primary products, 154–56; USDA definition of, 63; and water and sanitation, 148–53

hunger hotspots, 177

hunger vulnerability diamonds, 52

Hunger Vulnerability Index: components of, 13, 51–52; definition of, 3; and environmental systems, 85; as a hunger indicator, 1, 11, 51–56; 188–90; and income inequality, 120–21; and land, 81–82; and literacy 86–87; and national income, 129; and natural disasters, 178–80; and road networks, 89; and war, 174–77; weaknesses of, 53, 55

Hurricane Katrina, 179

IMF: effects of policies on hunger, 107

income distribution, and hunger, 120–21, 184

income inequality, and hunger vulnerability, 120–21

India: anemia in, 67; growth failure in, 68; hunger in, 21; Integrated Child Development Program, 181; malnutrition in, 67–68; political freedoms and famine, 115; poverty, 134; water, 150–51

inequality: gender 120–24; income, 120–21; and trade, 154

infrastructure, 88

institutions, 107; and hunger vulnerability 4; international financial, 111

Intergovernmental Panel on Climate Change (IPCC), 94–95

international trade, and hunger 154–56

Internet, digital divide, 101–2

iodine deficiency, 29–30

land: and hunger, 81–82, 184; utilization in Zimbabwe, 79

landlessness, in Brazil, 80

Latin America: coffee, 160, 161; HVI, 56; income distribution, 120; land, 80; obesity, 34, 36

life-expectancy, 122

literacy, and hunger 184; 86–87

Lorenz curve, 119

malaria, 143–44

malnutrition: definition of, 3; double burden of, 33

Malthus, Thomas, 75, 78

map: reading 7; projections, 7

Mexico: food protests in, 6–7; malnutrition in, 68–69; obesity in, 68–69; PROGRESA, 181

micronutrient deficiency: definition, 3; malnutrition, 26–31

Millennium Development Goals (MDG): and poverty, 45, 55, 131; and hunger 2, 13, 21, 182; and safe water, 148

Multicentre Growth Reference Study, 38

National Center for Health Statistics, 32

natural disasters: in China, 179; and hunger vulnerability, 179; Indian Ocean tsunami, 178; and food price shocks, 178–80; in the United States, 179

national income: and hunger vulnerability, 127–30, 184; and undernutrition, 130; and poverty rate, 130

neocolonialism, 109

New Rice for Africa (NERICA), 104

Nigeria, poverty in, 134

nutrition: individual, 3; transition, 33

obesity: in developing countries, 33–35; in Mexico, 68–69; in the United States, 33; in urban areas, 34

overnutrition, definition of, 3

overpopulation, 75

overweight. See obesity

Oxfam America, and global food crisis, 2

political freedoms, and hunger vulnerability, 115–17, 184

population growth, and hunger, 75–78, 184

Poverty and Hunger Index, 54

poverty: $2/day, 43–50; in Bangladesh, 134; in China, 131; extreme ($1.25/day), 44–46, 131–34; food availability and, 47–49; and growth failure, 46–47, 49; and hunger, 4, 51; household-level, 5; in India, 125–34; in Nigeria, 134; in the United States, 64–65

poverty line, 43

power relations, and hunger vulnerability, 5

prevalence of undernourishment (POU): as an indicator of hunger, 13, 19–25; useful functions of, 22; weakness of, 20–21

primary products: and hunger, 110; and international trade, 154–56; and terms of trade, 157

Programma de Alimentacíon Primaria (PAP), 59

purchasing power parity (PPP): and calculating poverty, 3–44, 131; and exchange rates, 128

refugees, 174

resource base, and hunger, 4, 91–93, 184

roads, and hunger, 88–90, 184

sanitation, 148–50

Sen, Amartya: and entitlements, 5; and political freedoms, 115

short-stature. See growth failure

Southeast Asia: colonialism, 109; HVI, 56; malaria, 143; trade, 156

structural adjustment, 107, 112

stunting. See growth failure

technology, and hunger vulnerability 4, 184

terms of trade: definition of, 157; and growth failure, 161; and hunger, 157–61; index, 157; and primary products, 157–59; shocks, 160–61

tied aid, 169–70

trade: and poverty, 155; in primary products, 154–55; food, 162–64

undernourishment: definition of, 3; number of people undernourished, 23, 182–83; prevalence of, 13, 182–83; rates of, 22 undernutrition: definition, 3

underweight, 37

United Nations High Commission on Refugees (UNHCR), 174

United States: agricultural subsidies, 158; arable land in, 83; biofuels, 163–64; consumption levels in, 78; cotton, 158; food aid, 169; food insecurity, 63–64; food stamps, 64–66; foreign aid, 168; hunger in, 63–66; Hurricane Katrina, 179–80; income distribution, 118, 120; land, 83; obesity in, 33; poverty in, 43, 63–64

United States Department of Agriculture (USDA): definition of hunger, 63; Food Stamp Program, 65–66; and healthy diets, 17

Via Campesina, and global food crisis, 2

Vietnam, and malaria, 143

Vitamin A deficiency, 26–30

war: in Darfur, Sudan, 176; and hunger, 174–77

wasting, 37

water: safe drinking, 7, 148–52; in Angola, 149; in Bolivia, 153; and hunger, 148; in India, 150–51; privatization, 153

women: and food security, 122–24; and nutrition, 182

World Bank: and global food crisis 2; and neocolonialism, 109; structural adjustment policies and hunger, 107

World Food Programme, 170–71

World Food Summit, 2, 182

World Health Organization (WHO): Global Database on Child Growth and Malnutrition, 37; nutrition surveys, 4

Zimbabwe: HIV/AIDS in, 78; population growth in, 76

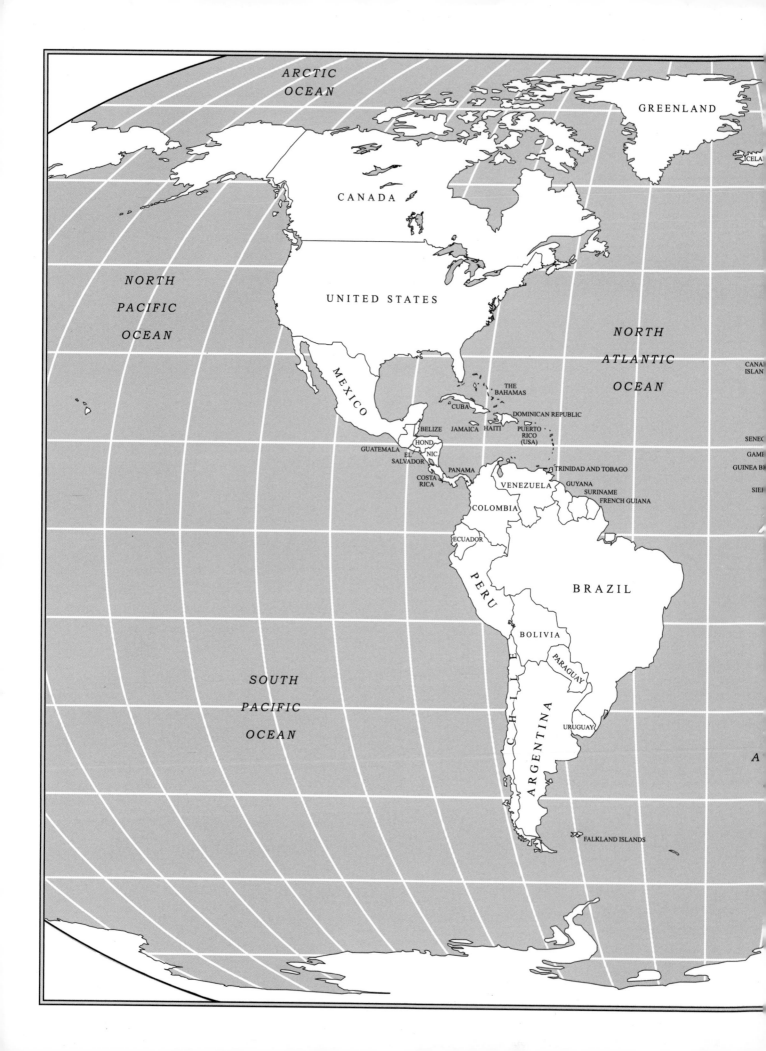